Library of
Davidson College

$4.95

E. J. LEMMON (1930—1966)

AMERICAN PHILOSOPHICAL QUARTERLY

MONOGRAPH SERIES

Edited by NICHOLAS RESCHER

THE "LEMMON NOTES"

AN INTRODUCTION TO MODAL LOGIC

by

E. J. LEMMON

In collaboration with Dana Scott

Edited by Krister Segerberg

Monograph No. 11　　　　　　　　Oxford, 1977

PUBLISHED BY BASIL BLACKWELL

© *American Philosophical Quarterly 1977*
ISBN 0 631 11550 1

160
L554l

77-6418
PRINTED IN ENGLAND
by J. W. Arrowsmith Ltd., Bristol, BS3 2NT

CONTENTS

Editor's Preface v
Historical Introduction 1
   Section 0 13
   Section 1 20
   Section 2 29
   Section 3 40
   Section 4 50
   Section 5 62
   Section 6 79
Bibliography 91

EDITOR'S PREFACE

This work is the printed version of the legendary Lemmon Notes. The circumstances surrounding them are special, and it may be well to try to say something about their way to publication.

During the 1960's California, "the Golden State," was probably the leading center in the world for the study of modal logic; in the literature reference is sometimes made to the "California school," although the expression has not gained currency. The El Dorado before others was the philosophy department at University of California, Los Angeles. Rudolf Carnap was emeritus and no longer active, but the atmosphere that he—and Hans Reichenbach?—had created still prevailed. Richard Montague was working his way towards what we now call Montague grammar, certainly an offshoot of modal logic. David Kaplan, while publishing sparsely, produced a steady stream of letters and dittoes for the inner circle. The department also contained, among others, Don Kalish and Montgomery Furth, both with interests in the philosophy of intensional logic. David Lewis lent further vigor to it when he joined it in 1966. A department of such brilliance would of course attract outstanding students; in fact, several of them have later gone on to do distinguished work in philosophical logic, for example, Nino Cocchiarella and Hans Kamp. Furthermore, visitors kept coming and going. In particular, Arthur Prior spent the fall of 1965 at U.C.L.A. as Visiting Flint Professor.

The number two modal logic arena in California at this time was Stanford, where Dana Scott was the dominating figure. Scott, a mathematician with an interest in philosophy, had been working with problems in tense and modal logic since the late 'fifties, and, like his Los Angeles colleagues Montague and Kaplan, had carried out much unpublished research. The philosophy department at Stanford was smaller than its counterpart at U.C.L.A., yet its atmosphere too held a certain excitement during those years. Patrick Suppes was chairman, Donald Davidson was still there, and Dagfinn Føllesdal and Jaakko Hintikka were commuting regularly from their alternative possible worlds in Scandinavia.

One prominent member of this bustling community of modal logic was E. J. Lemmon. Lemmon had received his education at

Oxford. His early work had been in classics; his doctoral thesis was on Plato. But later his interests had changed, perhaps due to Arthur Prior's influence. Although Lemmon maintained a keen interest in several departments of philosophy—ethics, for example—there is no doubt that as a mature philosopher his main interest was in logic, and in modal logic especially. His publications—some twenty papers, two books, and numerous reviews—are all in or not very far from logic.

At Oxford Lemmon had been research fellow in Magdalen College 1955–57 and tutorial fellow in Trinity College from 1957. The spring term of 1961 he spent as visiting professor in the University of Texas, Austin. Presumably he liked this American experience, or he was attracted by the favorable climate for modal logic in southern California, for in 1963 he came to the U.S. for good, joining the faculty at Claremont Graduate School near Los Angeles. He was about to take up a position in the University of California, Irvine, when, because of heart failure, he died on July 29, 1966. At his death he was 36 years old.

When the present editor became a student of Scott in 1965, the collaboration between Lemmon and Scott was well under way. Their objective was to write a monograph whose primary aim, in Lemmon's own words, was "the suitable deployment of theoretical semantics" for intensional languages. Both in Los Angeles and Stanford there was a tendency to avoid using the term "modal logic" in the inclusive sense it is often used—as, for example, in this preface. This reluctance is documented in Montague's writings, and Lemmon and Scott shared it. The working title of their proposed monograph was *Intensional logic*—a conscious and rather pointed aberration from current usage, certainly striking in a work that aspired to be the Bourbaki of modal logic. The title of the present work, *Introduction to modal logic*, due to Nicholas Rescher, is no doubt in better agreement with the tradition they wanted to oppose, and for this reason perhaps more informative.

There are various indications that the book was planned to consist of a historical introduction and five chapters. Of these the first three chapters would deal with propositional logic. Chapter I would lay down the basic theory for normal modal logic. It seems likely that Chapters II and III would be devoted to topics in deontic, epistemic and tense logic, and that they would also include discussions of nonnormal logics, the so-called neighborhood semantics and certain other unorthodox modellings (cf. pp. 12, 19, 23, 30, 42,

61, 69, 73, 78 and 82 below). Chapters IV and V would then deal with predicate logic (cf. the opening sentence of *Section 0* on p. 13). It is of course impossible to know exactly what the book was meant to contain, and probably the collaborators were not quite sure themselves.

The present work represents the only part of Lemmon's and Scott's project that ever reached a degree of completion. Actually it is a draft of the historical introduction and Chapter I. The manuscript, written in longhand by Lemmon, is dated July 26, 1966. After Lemmon's death, Scott arranged to have the manuscript typed and dittographed. The title page of the dittoed edition bears this inscription: "INTENSIONAL LOGIC by E. J. Lemmon and Dana Scott. Preliminary draft of Initial Chapters by E. J. Lemmon. July 1966." Since then the Lemmon Notes have spread in ever widening circles, generation after generation of xerox copies of monotonically decreasing legibility. Considering their underground mode of publication, they have had a remarkable impact; in contemporary modal logic they are an oft-quoted reference.

Lemmon provided no titles for the seven sections that make up his Chapter I. Briefly, what they contain is this. *Section 0* is devoted to basic syntax of classical propositional logic. Lindenbaum's Lemma is proved. *Section 1* introduces the slightly modified Kripke type possible-worlds-semantics that Lemmon and Scott preferred. *Section 2* is devoted to a first discussion of completeness: a Henkin type completeness proof for the smallest normal modal logic K is given, but also another kind of completeness is contemplated, depending on a novel notion of "strong satisfiability." In *Section 3*, K is shown to be decidable by two different methods, one syntactical and one semantical. *Section 4* presents a very general completeness theorem which covers all normal logics definable in terms of schemata of type $\Diamond^m \Box^n \mathbf{A} \to \Box^p \Diamond^q \mathbf{A}$. It turns out that a great many of the systems in the literature are of this kind, not only the bestknown ones—the Gödel/Feys/von Wright system T, Lewis's $S4$ and $S5$—but many others as well. This very general result is then, in its turn, very considerably generalized in *Section 5*, where there is also a discussion of the possible limitations to the Henkin method in modal logic. *Section 6*, finally, is devoted to a closer look at particular systems. The two main topics discussed here are decidability and the number of nonequivalent irreducible modalities.

Roughly, the interdependence of the seven sections is indicated by the following chart:

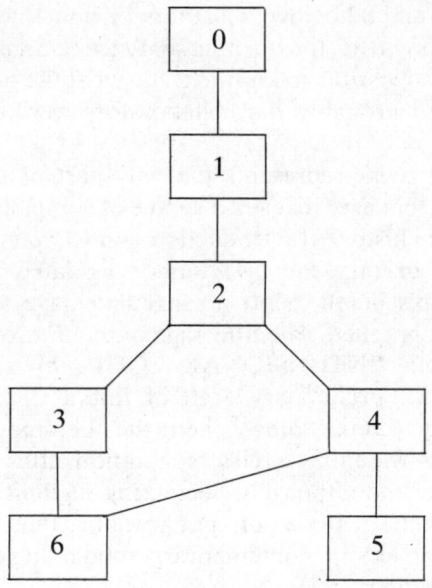

Perhaps it is the approach and the spirit that make the Lemmon Notes into the modal logical classic it already is. But there is also a wealth of new methods and new results. Probably the most important of the firsts is the introduction of Henkin type completeness proofs into modal logic. The same idea has also occurred, independently, to some other authors, for example David Makinson and M. J. Cresswell, but only Lemmon and Scott seem to have realized the great potentialities for exploiting this idea systematically, as they set out to do in *Sections 4* and *5*. Moreover, the semantical decision method described in *Section 3* and further elaborated in *Section 6* is the beginning of what is now called the filtration technique. These two elements in particular have influenced the course of modal logic. Of the many other interesting features of the Lemmon Notes at least one must be mentioned here, viz., the notion of completeness introduced at the end of *Section 2*. This whole discussion is very suggestive in light of later work by David Makinson and S. K. Thomason; it really anticipates Thomason's results on what he calls first order models.

Inevitably it will be asked what the Lemmon Notes owe to Lemmon and what they owe to Scott. The answer to that question, as far as the editor knows, is that the Lemmon Notes were very much the result of real collaboration and that it is difficult, or impossible, to sort out the credit due to each man. The fact that it was Lemmon who wrote the draft means of course that he has left his characteristic imprint on it; many details of presentation must be his. It is also tempting to conjecture that most of the historical introduction is his work. But as far as the rest goes it seems futile to try to divide the credit; Dana Scott, when asked, insists that it is. In this connection, it may be of interest to append the following information that Scott has conveyed to the editor. Lemmon and Scott had met many times and had carried on an extensive correspondence exchanging proofs before Lemmon had ever started writing his version. At one point, when it had become clear that publication of the planned work would require much more time than Lemmon had hoped, he inquired whether Scott would think it proper for him to publish separately, under his own name, the results that were obtained during the collaboration but that were due to him. This Scott encouraged. Evidently, the material Lemmon referred to on this occasion was what went into the four important papers that Lemmon published in *Notre Dame journal of formal logic* for 1965 and 1966 and *The journal of symbolic logic* for 1966.

There are three respects in which the Lemmon Notes are incomplete. First, as we have explained at length, they are but a fragment, if a substantial one, of the planned *Intensional logic*. Second, Lemmon had no notion that his manuscript would be published in its present form. Not surprisingly, a careful reader will discover minor flaws in the exposition, flaws of a kind never to be found in the polished writings that Lemmon himself saw through the press. Third, the bibliography is missing.

From an editorial point of view, the major respect in which Lemmon's draft is incomplete is the third. Like so many authors Lemmon would enter into his handwritten text references of type "Prior [ ]," having in mind a certain work whose identity was to be made explicit at a later stage by inserting within the brackets an appropriate symbol. Evidently the bibliography was never put on paper. *It must be emphasized that the bibliography appearing here is a reconstruction.*

Even though Lemmon did not have immediate publication in mind, his draft, as everything he wrote, is eminently readable. Therefore it was decided to publish it as it is, with as few changes as possible. Accordingly it is in only a small number of places that the text has been corrected. In particular, no assertion or view that Lemmon expressed has been altered. In places where editorial intervention has seemed desirable, it has taken the form of footnotes; in fact, all the footnotes appearing in this edition are the editor's. Thus the vast majority of changes are trivial. They concern punctuation, spelling and typography, and they are always made in order to increase clarity and consistency without infringing upon Lemmon's taste.

The editor wishes to thank Dana Scott for entrusting to him the task of making Lemmon's manuscript publishable. Thanks are also due to Mrs. Lee Lemmon and to J. F. A. K. van Benthem, Roderick M. Chisholm, David Fowler, Bengt Hansson, Jaakko Hintikka, Stig Kanger, Steven T. Kuhn, Simo Knuuttila, Mary Prior, W. V. O. Quine, Arthur Skidmore, and G. H. von Wright for their kind assistance in connection with the work on this edition, in particular to Kuhn for his help in checking the final text against the handwritten original.

Finally, on behalf of all present and future students of modal logic, the editor should like to thank Nicholas Rescher for his willingness to publish the Lemmon Notes in the American Philosophical Quarterly Monograph Series. Without his efficient cooperation their publication would have been even further delayed.

Lawrence, Kansas *Krister Segerberg*
Thanksgiving Day, 1976

Historical Introduction

The history of modal logic, or (as we prefer to call it) intensional logic, is long but easily surveyed. It begins, as does the history of logic itself, with Aristotle, who devoted two chapters (12 and 13) of the *De Interpretatione* to a study of the logical interconnections between the necessary, the impossible, the possible, and the permitted. Much of the discussion is confused: Aristotle has difficulty seeing how properly to negate modal sentences (should the negation of 'possible to be' be 'not possible to be' or 'possible not to be'?); and, once he has this straight, he has difficulty setting out the implications between different modal sentences. Nonetheless, the outcome of his discussion is a remarkably correct set of implications.

Let us adopt an informal notation to express some of these insights. For proposition A let $\Box A$ mean that A is necessary and $\Diamond A$ mean that A is possible. Then Aristotle is quite clear that $\Box A$ and $-\Box A$, $\Diamond A$ and $-\Diamond A$ are contradictory pairs, whilst $\Box A$ and $\Box - A$, $\Diamond A$ and $\Diamond - A$ are not. He is also clear that the necessary implies the possible:

$$\Box A \to \Diamond A. \qquad (1)$$

He is inclined at first, however, also to accept the principle

$$\Diamond A \to \Diamond - A \qquad (2)$$

(what is possible is also possibly not), but sees soon enough that (1) and (2) together yield the absurdity that what is necessary is also possibly not. Clearly what is required to resolve this paradox is a distinction between two senses of 'possible', for one of which (possibility proper) (1) holds but not (2) and for the other of which (contingency, say) (2) holds but not (1). And Aristotle comes close, at the end of Chapter 13, to making just this distinction. He also in this chapter formulates two other modal principles, that what is of necessity is in actuality and that what is in actuality is also capable of being, which we may represent

$$\Box A \to A, \qquad (3)$$

$$A \to \Diamond A. \qquad (4)$$

In fact (1) is a simple consequence of (3) and (4), though Aristotle does not point this out.

As for the impossibility of **A** (I**A**), Aristotle (again after a struggle) comes to see that I**A** is equivalent to $\Box - \mathbf{A}$ (he says that $\Box$ and I have the same force, but contrary subjects):

$$I\mathbf{A} \leftrightarrow \Box - \mathbf{A}. \tag{5}$$

He comes close to recognizing that I**A** is equivalent also to $-\Diamond \mathbf{A}$:

$$I\mathbf{A} \leftrightarrow -\Diamond \mathbf{A}, \tag{6}$$

but here his insights are slightly marred by his tendency to accept (2).

In the *Prior Analytics*, after his development of the assertoric syllogism, Aristotle treats modal syllogisms, syllogisms with modalized premisses and conclusion (*Pr. Anal.* I, cc. 3, 8–22). In this sketch, we shall not concern ourselves with this aspect of the history of modalities (for modern treatments see Sir David Ross [1], A. Becker [1], Lukasiewicz [1], and Storrs McCall [1]). There are two aspects of his treatment, however, that we do need to note. Firstly, Aristotle is careful to distinguish two senses of 'permitted to be' (essentially our $\Diamond \mathbf{A}$); in its primary sense it means 'that which is not necessary and which, if assumed to be the case, has no impossible consequences' (32a18–20); in its second sense, he says, the necessary *is* permitted, so that we must assume that this sense is given simply by 'that which, if assumed to be the case, has no impossible consequences' (32a20–21). Now in modern terms that which, if assumed, has no impossible consequences is simply equivalent to the not impossible. Writing, therefore, Q**A** for the first sense and preserving $\Diamond \mathbf{A}$ for the second, we have

$$Q\mathbf{A} \leftrightarrow -\Box \mathbf{A} \wedge -I\mathbf{A}, \tag{7}$$

$$\Diamond \mathbf{A} \leftrightarrow -I\mathbf{A}. \tag{8}$$

In the light of this contrast, the uncertainty in the *De Interpretatione* is finally resolved. (8), of course, squares exactly with (6). (7) on the other hand squares with (2) in the sense that, if in (2) we replace $\Diamond$ by Q, we obtain

$$Q\mathbf{A} \rightarrow Q - \mathbf{A}, \tag{9}$$

which is in fact a consequence of (7) and (5). (Given Q**A**, $-\Box \mathbf{A}$ and $-I\mathbf{A}$ by (7), whence $-I - \mathbf{A}$ and $-\Box - \mathbf{A}$ by (5), whence $Q - \mathbf{A}$ again by (7).) With (9) replacing (2), all Aristotle's theses so far are acceptable by modern standards. A further consequence of (7) and (8) is

$$Q\mathbf{A} \rightarrow \Diamond \mathbf{A}; \tag{10}$$

Aristotle appears to use this in his modal syllogistic, though nowhere explicitly formulates it. The converse of (10), of course, he recognizes as false: a necessary proposition is possible by (1), but not contingent by (7).

Secondly, Aristotle writes as follows (34a5–24): 'We must first say that, if when A is it is necessary that B should be, then if A is possible it is also necessary that B should be possible.... So that if one calls the premisses [of a syllogism] A and the conclusion B, it will result not only that B will be necessary if A is necessary, but also that it will be possible if A is possible'. It is natural to find in these words two important insights: (a) that a necessary conditional with possible antecedent has a possible consequent; (b) that a necessary conditional with necessary antecedent has a necessary consequent. In our informal notation, these are

$$\Box(A \to B) \to (\Diamond A \to \Diamond B), \tag{11}$$

$$\Box(A \to B) \to (\Box A \to \Box B). \tag{12}$$

Aristotle's words in the first quoted sentence support a rather different interpretation, namely belief in the principles

$$\Box(A \to B) \to \Box(\Diamond A \to \Diamond B), \tag{13}$$

$$\Box(A \to B) \to \Box(\Box A \to \Box B). \tag{14}$$

We shall see later that there is a crucial difference between (11) and (12) on the one hand and (13) and (14) on the other.

The Megarians and Stoics, in addition to their well-known work on the propositional calculus (see Mates [1]), also developed various theories concerning modality. We learn from Boethius (*Commentaries on Aristotle's De Interpretatione*, second part, second edition, ed. Meiser, p. 234 (Leipzig: Teubner, 1880))[1] that 'Diodorus defines the possible as that which either is or will be; the impossible as that which, being false, will not be true; the necessary as that which, being true, will not be false; and the non-necessary as that which either is already or will be false.' This is of great interest, since it suggests a connection between the modalities and temporal notions (what will be the case). This connection is supported by recent work on the logic of time (see Prior [1]), and we shall explore it ourselves in a later section. Various other Megarian and Stoic

---

[1] The precise title of this work, which is written in Latin, is *Ancii Manlii Severini Boetii commentarii in librum Aristotelis Περὶ Ἑρμηνείας recensuit Carolus Meiser. Pars posterior. Secundam editionem et indices continens*. No credit is given for the translation into English of the passage quoted, and it may be due to Lemmon himself.

attempts to define the necessary and the possible are reported in many writers of antiquity; there is no evidence, however, that any major *formal* advance on what had been achieved by Aristotle was made. For further details see W. and M. Kneale [1] and Prior [2].

In the middle ages, a lot of attention was given by logicians to modal notions. Apart from work on the modal syllogisms and the somewhat confusing contrast between modalities *de re* and modalities *de dicto* (compare Aquinas, *De Modalibus*[2]), with which we are not here concerned, we find in various authors careful statements of Aristotelian modal principles and variants of them. Some principles are assigned Latin tags; thus (3) becomes *A necesse esse ad esse valet consequentia*, and (4) *Ab esse ad posse valet consequentia*. In Ockham and others, we further find *rejections* of principles, such as: the impossible does not follow from the possible, the contingent does not follow from the necessary. Pseudo-Scotus adds the variations

$$-\Box A \leftrightarrow \Diamond - A, \tag{15}$$

$$-\Diamond A \leftrightarrow \Box - A, \tag{16}$$

of which (16) is implicit in Aristotle's (5) and (6). Indeed, Pseudo-Scotus is particularly rich in modal ideas. In addition to 'necessary' and 'possible', he considers modes such as 'it is doubtful that', 'it is known that', 'it is believed that', 'it is wished that', and points out logical similarities between them and the usual modalities, thus anticipating very recent work on so-called epistemic logics and others which stems perhaps from von Wright [2]; we shall revert to this later. Finally, he notes both (11) and (12) and extends them to other modalities. (See W. and M. Kneale [1], Bocheński [1]; variants on (11) and (12) may be found rather earlier in Abélard, another notable medieval logician.)

Perhaps even more remarkable are medieval anticipations of the so-called paradoxes of strict implication. The Stoic Philo had suggested in effect that 'if ... then ...' should be analysed by means of a truth-table as material implication. By contrast, Diodorus appears to have suggested that 'if ... then ...' should be analysed as something like a *necessary* material implication. To represent this, we might agree that $\rightarrow$ stands for material implication, and now define $A \Rightarrow B$ as $\Box(A \rightarrow B)$, so that $\Rightarrow$ stands for Diodorus' 'if ... then ...'. We find many medieval writers insisting that any true conditional must be necessarily true, roughly on the grounds

---

[2] For the text with a lengthy commentary, see I. M. Bocheński, 'Sancti Thomae Aquinatis De Modalibus Opusculum et Doctrina', *Angelicum*, vol. 17 (1940), pp. 180–218.

that to affirm if **A** then **B** is to affirm that it is *not possible* for **A** to be the case *without* **B**. This is related closely to the principle

$$\Box(\mathbf{A} \to \mathbf{B}) \leftrightarrow -\Diamond(\mathbf{A} \wedge -\mathbf{B}) \tag{17}$$

(where we use $\wedge$ for conjunction). Abélard, however, who did believe that conditionals if true are necessarily so, was worried by (17) because it commits one to the truth of 'if Socrates is a stone, then Socrates is an ass' simply given the impossibility of Socrates' being a stone. For if Socrates' being a stone is impossible, then it is impossible for him to be a stone without being an ass. Abélard has seen, in effect, that, if **A** is impossible, then $\mathbf{A} \Rightarrow \mathbf{B}$ is true for *any* proposition **B**:

$$-\Diamond \mathbf{A} \to (\mathbf{A} \Rightarrow \mathbf{B}). \tag{18}$$

Later, Kilwardby notes the other side of the same coin; if **B** is necessary, then it is not possible to have **A** without **B**, since it is not possible to have $-\mathbf{B}$, so that again by (17) $\Box(\mathbf{A} \to \mathbf{B})$. This gives

$$\Box \mathbf{B} \to (\mathbf{A} \Rightarrow \mathbf{B}). \tag{19}$$

Similar arguments are found in Pseudo-Scotus, and Ockham gives explicitly the principles: (a) anything whatsoever follows from the impossible; (b) the necessary follows from anything whatsoever. Here (a) corresponds to (18) and (b) to (19). (For more detailed treatment of these matters, see W. and M. Kneale [1].)

From the close of the middle ages until the nineteenth century, there is virtually no trace of attention paid to problems in modal logic. The modalities appear, of course, in Kant's Table of the Categories, but there is no sign that Kant was interested in their formal properties. Kant's distinction between analytic and synthetic judgments, however, which has its roots in earlier distinctions between truths of reason and truths of fact, is worth mentioning since it is now clear that the notion of analyticity is an intensional notion in a sense we shall try to make plain later. Frege's view in the *Begriffsschrift* is that modal notions such as 'must' are really epistemic and convey a reference to human knowledge; as such, they are irrelevant to pure logic. On the other hand, what Frege says about oblique contexts in 'Sense and Reference' is directly relevant to problems of intensionality. Contemporary attention to the formal properties of modal terms really begins with the work of C. I. Lewis, though there are anticipations in Hugh MacColl (*Symbolic Logic and its Applications*, 1906).

Lewis' starting point is the paradoxes of material implication, as they appear in *Principia Mathematica*, and his concern is to develop a formal system which will contain an implication lacking these paradoxes. This new implication he calls *strict* implication, and in the *Survey of Symbolic Logic* (1918) it is defined in terms of conjunction, negation, and a primitive operation of impossibility :

$$A \Rightarrow B =_{df} (|A \wedge -B),$$

a style of definition clearly reminiscent of (17). Lewis, like his medieval predecessors saw that this definition led to the *paradoxes of strict implication* (18) and (19), but he regards these as 'inescapable consequences of logical principles which are in everyday use' (Lewis and Langford [1], p. 175). They have nonetheless been the topic of much subsequent dispute.

In the later *Symbolic Logic*, jointly written with C. H. Langford, we find a clearer exposition of Lewis' formal work on the modalities.[3] Here $\Diamond$ is taken as a primitive operation in place of | —clearly a better choice—and strict implication defined

$$A \Rightarrow B =_{df} -\Diamond(A \wedge -B).$$

In the course of discovering the formal properties of strict implication, many purely modal principles are enunciated. In fact, this work describes five distinct systems of modal logic, *S1–S5*, nowadays usually called the Lewis systems. Lewis first presents *S1*, and extends his discussion successively to the stronger *S2* and the stronger still *S3* (equivalent to the system originally presented in the *Survey of Symbolic Logic*). *S4* and *S5* are discussed in Appendix II of *Symbolic Logic*. All these systems have been greatly discussed in the modal literature, and we shall have much to do with them in the sequel.

A little before Lewis and Langford [1] appeared Gödel [1], in which a modal logic was suggested in which the usual propositional calculus was to be taken as a base and certain modal axioms and rules added. Gödel in fact proposed taking (3) and (12) as axioms, together with

$$\Box A \rightarrow \Box\Box A, \qquad (20)$$

and adding a rule of inference: from $\vdash A$ to infer $\vdash \Box A$. He further claimed the equivalence of this system to the Lewis system *S4*, though a proof of this equivalence was not published until McKinsey and Tarski [1]. Although Lewis himself was aware that all five of his systems contained the classical propositional calculus, he did not

---

[3] Notice that Lewis and Langford [1] and *Symbolic Logic* are identical.

use this fact in his formulation of them but preferred rather clumsy axiom sets stated in terms of strict implication. The first formulation of the systems *S1*–*S3* in the Gödel manner, with the usual propositional calculus separated out as a base, would appear to be in Lemmon [1]. It in general proves simpler in formulating systems to do this, and also to imitate Gödel in taking $\Box$ rather than $\Diamond$ as primitive. Then strict implication can quite easily be defined in terms of material implication:

$$\mathbf{A} \Rightarrow \mathbf{B} =_{df} \Box(\mathbf{A} \rightarrow \mathbf{B}),$$

and (17) correspondingly turns up as a theorem. Implicit in both Aristotle and later writers are the equivalences

$$\Diamond \mathbf{A} \leftrightarrow -\Box - \mathbf{A}, \qquad (21)$$

$$\Box \mathbf{A} \leftrightarrow -\Diamond - \mathbf{A}, \qquad (22)$$

so that with either $\Box$ or $\Diamond$ as primitive the other may readily be defined. Other modal notions such as I or Q, may then, if desired, be defined in terms of these (compare (5), (6), and (7)).

Though several modal systems were well-defined as axiomatic systems by 1932, still not much was known about them (e.g., *S3* was not shown to be a *distinct* system from *S2* until Parry [1]). A basic article from this period is Parry [2], from which it is clear that there are fundamental differences between *S3* and *S2*. For example, *S3* offers (13) and (14) as theorems whilst *S2* does not. Another important contribution was Feys [1], in which the system *T* (what Feys called 'logique t') was first described—essentially the Gödel postulates mentioned earlier less (20). This system, unnoticed by Lewis, turned out to be stronger than *S2*, weaker than *S4*, and independent of *S3*. It is equivalent to the system *M* of von Wright [2].

It is impossible here to mention in detail developments after 1940; a good survey is Prior [2] or Feys [2]. In any case, many points will be discussed in the sequel. But McKinsey [2] should be noted, in which a characterization of the systems *S2* and *S4* from an abstract algebraic standpoint was given, as a result of which both systems were shown to be decidable. Similar algebraic techniques have been applied subsequently by various writers to obtain results concerning modal logics. McKinsey [3] offers what he calls a syntactical treatment of modal logic, and we shall say more of this later.[4] Carnap [1] and Barcan [1] offer perhaps the earliest systematic attempt to

---

[4] This may refer to the discussion at the end of Section 2. At least there would seem to be a connection between the material discussed there and McKinsey's work.

combine modal operations with quantifiers. Ohnishi and Matsumoto [1] produced the first Gentzen-type formulation of certain modal logics (though natural deduction techniques are in Fitch [1]).[5] Many variant brands of modal logic have also been produced: epistemic logic, doxastic logic, deontic logic, tense logic (see von Wright [2], Prior [2] and [3]). Finally, and quite recently, Kripke [1] and [2] gives the first clearly semantical account of modal logic (with the exception of Carnap [2]), combined with decision procedures based on semantic tableaux of a Beth type. It is in the spirit of Kripke-style semantics that this book is written.

Since it is our view that modal logic is really the study of intensional as opposed to extensional contexts, we must go back and trace briefly the history of this aspect of our field. The medieval term *intentio* was originally employed as a translation of the Arabic *ma'na*, a form in the soul identified with a meaning or notion, and meant throughout medieval epistemology a natural sign in the soul. Later the *Port Royal Logic* distinguished between the *comprehension* and *extension* of a general term in something of the way in which Mill later distinguished connotation and denotation: whilst the extension is the set of things to which the term applies, its comprehension is the set of attributes which it implies. In the nineteenth century, Sir William Hamilton replaced 'comprehension' by 'intension', faultily spelling the word with an 's' by analogy with 'extension'. Since then, the term 'intentionality' has gone one way, via Brentano to Chisholm [1], [2] and the word 'intensionality' another via Carnap to Quine.[6] It is the latter term in which we are most interested, though there is no doubt that it is intimately connected with the former.

The problem raised by intensionality may be briefly stated by going back to the Leibnizian: *eadem sunt quorum unum potest substitui alteri salva veritate*—those things are the same of which one may be substituted for the other with preservation of truth.

---

[5] Lemmon's formulation might suggest that Ohnishi and Matsumoto were the first to study Gentzen type formulations of modal logic. Actually they were not. Anyone interested in the history of this particular subject should consult at least the following works: Haskell B. Curry, 'The elimination theorem when modality is present', *The journal of symbolic logic*, vol. 17 (1952), pp. 249–265; Stig Kanger, *Provability in logic* (Stockholm: Almqvist & Wiksell, 1957); J. Ridder, 'Die Gentzenschen Schlussverfahren in modalen Aussagenlogiken', *Indagationes mathematicae*, vol. 17 (1955), pp. 163–177, 270–276.

[6] According to the manuscript, Lemmon had intended only one reference here, presumably to one of the reconstructed Chisholm references.

Sometimes called the substitutivity of identity, sometimes the principle of extensionality, this law turns up roughly as follows in contemporary treatments of identity:

$$a = b \to (\cdots a \cdots \leftrightarrow \cdots b \cdots), \tag{23}$$

where $\cdots a \cdots$ is a context containing occurrences of the name **a**, and $\cdots b \cdots$ is the same context in which one or more occurrences of **a** has been replaced by **b**: if **a** is **b**, then what goes for **a** goes for **b** and vice versa. In the ordinary propositional calculus, we have a similar principle of the *substitutivity of material equivalents*:

$$(A \leftrightarrow B) \to (\cdots A \cdots \leftrightarrow \cdots B \cdots). \tag{24}$$

Indeed, on the view of Frege in 'Sense and Reference', (24) is merely a special case of (23), since $A \leftrightarrow B$ represents an *identity of truth-value* between **A** and **B**. Towards the end of 'Sense and Reference', Frege draws attention to cases of (real or apparent) failure of (23) and (24), and calls them *oblique* contexts—nowadays they are usually called *intensional* or *referentially opaque* (Quine [1]).[7] We may illustrate the point here by reference to modal examples, though 'it is known that', 'it is believed that', etc., provide similar cases. The morning star is the evening star ($a = b$), but though it is necessary that the morning star be the morning star ($\Box(a=a)$) it is not necessary that the morning star be the evening star ($-\Box(a=b)$). Indeed, the point may be put more strongly: if (23) were indiscriminately applied to *all* modal contexts, then given $\Box(a=a)$ we could demonstrate that *all* true identities are necessarily true. Thus (23) yields

$$a = b \to (\Box(a=a) \to \Box(a=b)), \tag{25}$$

whence by propositional calculus

$$a = b \to \Box(a=b). \tag{26}$$

A similar paradox can readily be generated from (24) if applied to modal contexts. Thus as a special case of (24)

$$(A \leftrightarrow B) \to (\Box A \leftrightarrow \Box B). \tag{27}$$

Now if **A** and **B** are both true, then they are materially equivalent. It is tautologous that

$$A \wedge B \to (A \leftrightarrow B) \tag{28}$$

---

[7] See also 'Reference and modality' in Quine's *From a logical point of view* (Cambridge, Mass.: Harvard University Press, 1953).

(27) and (28) yield by propositional calculus

$$A \wedge \Box A \to (B \to \Box B). \tag{29}$$

By (3), $\Box A \to A$, so that (29) yields

$$\Box A \to (B \to \Box B). \tag{30}$$

We have only to pick for **A** any necessary truth, say $a = a$, to conclude

$$B \to \Box B \tag{31}$$

—any proposition **B** which is true is true of necessity. It is clear that, unless these arguments are somehow blocked, the enterprise of modal logic will be at best trivial (in that (31) and (3) together collapse necessity into truth), at worst absurd.

An attempt to formalize Frege's own way out of these difficulties may be found in Church [1]. Meanwhile Quine [1] and [2] has subjected modal logic to severe attack,[8] principally on the grounds that it is committed to the existence of obscure intensional entities (Fregean senses, propositions, meanings) whose status cannot really be clarified, or at least committed to making sense of 'intensional' notions such as synonymy, analyticity, and the like, which Quine despairs of explicating. More recently (Quine [3], Chapter 6), he has argued that there is simply *no need* for intensional objects, that contexts which appear to demand them can be defined or analysed away. It is not our plan at this stage to answer these charges; but it is certainly our hope that the giving of suitable 'extensional' semantics for modal logics will remove a good deal of the mistrust of them which Quine and others have expressed. Thus we do not *deny* the *thesis of extensionality* so called, that any utterance made in an intensional language can be translated without loss into an extensional language. But we would deny the *need* for such a translation.

If we return to the derivation of the paradoxical conclusions (26) and (31), we may observe three ways of escape. (i) We may deny that a proper name **a** or sentence **S** has the same reference within a modal context as it has outside it; this is the Frege–Church escape-route. (ii) We may deny that any sentence expresses a necessary truth, i.e., assert that no sentence of the form $\Box A$ is true.[9] This is in a way Quine's solution, except that he quite sensibly draws the

---

[8] See also Quine's 'Three grades of modal involvement' in *Actes du XIème Congrès International de Philosophie*, vol. 14, pp. 65–81 (Amsterdam: North-Holland, and Louvain: Nauwelaerts, 1953), reprinted in idem, *The ways of paradox and other essays* (New York: Random House, 1966).

[9] The word 'assert' does not occur in the manuscript.

further consequence that modal logic is in this light a waste of time. It is also the solution of Lukasiewicz [1], who accepts (27) and even (30) but rejects (31); Lukasiewicz does not, however, draw the conclusion that modal logic is a waste of time.[10] (iii) We may deny (27), and so the full applicability of principles of extensionality such as (23) and (24). This is the course followed in most systems of modal logic as we shall see in the sequel.

The rejection of (27) on the grounds that it yields the unacceptable (31) if there are necessary truths carries with it the further rejection of

$$(A \to B) \to (\Box A \to \Box B). \tag{32}$$

Given (32), (27) follows by propositional calculus reasoning. And (32) yields (30) (and so (31), given a necessary truth) quite directly *via* the tautology $B \to (A \to B)$. It will be quite typical, therefore, for a modal system to contain (12) as a theorem but to lack the far stronger (32).

Despite the rejection of (27) and (32) in most systems of modal logic, there are closely related *rules of derivation* that *are* usually accepted. Thus corresponding to (32) there is the rule, to be called in the sequel *RM*:

RM: if $\vdash A \to B$ then $\vdash \Box A \to \Box B$.

It may well be that if $A \to B$ is a *logical* truth then so is $\Box A \to \Box B$. And corresponding to (27) there is the rule, to be called *RE*:

RE: if $\vdash A \leftrightarrow B$ then $\vdash \Box A \leftrightarrow \Box B$.

It may well be that, if A is *logically* equivalent to B, then $\Box A$ is *logically* equivalent to $\Box B$. (The situation here may be compared to that in predicate calculus, where $A(x) \to (\forall x) A(x)$ is not in general a theorem but from $\vdash A(x)$ we may conclude $\vdash (\forall x) A(x)$.)

---

[10] It is not the well-known 3-valued logic of Lukasiewicz that is referred to here, for, although it validates (27) and rejects (31), that logic does not validate (30). (See Lukasiewicz's 'Philosophische Bemerkungen zu mehrwertigen Systemen des Aussagenkalküls' in *Comptes rendues des séances de la Societé des Sciences et des Lettres de Varsovie*, cl. iii, vol. 23 (1930), pp. 51–77, translated by H. Weber and reprinted in *Polish logic 1920–1939*, edited by Storrs McCall (Oxford: Clarendon Press, 1967).) In his later work [1] Lukasiewicz says (§ 49) of his 1930 system that it 'does not satisfy all our intuitions concerning modalities' and that it should be replaced by a certain 4-valued logic. It seems to be the latter Lemmon has in mind—at least it validates (27) and (30) and rejects (31).

There are systems of modal logic that do not provide *RE*—Lewis' *S2* and *S3*, as we shall see, are cases in point.[11] Intuitively, the rule appears to have a rather fundamental character, since it enables us to distinguish certain intensional notions from other, perhaps more deeply intensional, ones. For example, if we are hoping that $\Box$ may be interpreted as 'it is analytic that', then *RE* looks like being correct. For if $\mathbf{A} \leftrightarrow \mathbf{B}$ is a logical truth, then $\mathbf{A}$ is analytic if and only if $\mathbf{B}$ is—however, 'analytic' may eventually be defined. On the other hand, if we have in mind *epistemic* interpretations of $\Box$ where it is to mean 'it is known that' (see, e.g., Hintikka [1]), then *RE* is quite implausible: $\mathbf{A} \leftrightarrow \mathbf{B}$ may be a logical truth such that $\mathbf{A}$ is known and $\mathbf{B}$ unknown. (*RM* is similarly implausible for this interpretation.) But these matters will become more amenable to discussion in the light of the formal development, to which we now proceed.

---

[11] *S2* and *S3* are not treated in this work. For analyses of those systems see, for example, Saul A. Kripke, 'Semantic analysis of modal logic II' in *The theory of models*, edited by J. W. Addison, L. Henkin and A. Tarski, pp. 206–220 (Amsterdam: North-Holland, 1965) or Lemmon's own paper [3].

Section 0

For the first three chapters, our main concern will be with *propositional* modal logics; extensions to quantifiers are made in the last two chapters.[12] At this initial level, we study formal sentences of a particular kind, which we shall call *modal sentences* (or usually just *sentences*). We give first, then, a definition of this notion.

By a *sentence letter* we understand one of the symbols

$$\mathbf{P}_0, \mathbf{P}_1, \ldots, \mathbf{P}_n, \ldots$$

which is to be taken as a list indefinitely continuable. We also introduce the constant symbol $\bot$ to be read 'the false', and reintroduce formally from the Introduction the symbols $\rightarrow$ and $\Box$, the first to represent material implication, and the second necessity. A modal sentence can be defined recursively as follows:

(i) a sentence letter is a modal sentence;
(ii) $\bot$ is a modal sentence;
(iii) if $\mathbf{A}$ and $\mathbf{B}$ are modal sentences, so is $(\mathbf{A} \rightarrow \mathbf{B})$;
(iv) if $\mathbf{A}$ is a modal sentence, so is $\Box \mathbf{A}$.

From now on, the letters $\mathbf{A}, \mathbf{B}, \ldots, \mathbf{A}_1, \ldots$ are used as *metalogical* variables, in contrast to their informal use in the Introduction, to range sometimes over symbols or sequences of symbols and sometimes over modal sentences in particular; exactly *how* they are being used will be clear from context, if their metalogical character is borne in mind.

Various other symbols are introduced by definitions, many of which should be familiar:

D1: $-\mathbf{A} = (\mathbf{A} \rightarrow \bot)$.

D2: $\mathbf{A} \vee \mathbf{B} = -\mathbf{A} \rightarrow \mathbf{B}$.

D3: $\mathbf{A} \wedge \mathbf{B} = -(-\mathbf{A} \vee -\mathbf{B})$.

D4: $\mathbf{A} \leftrightarrow \mathbf{B} = (\mathbf{A} \rightarrow \mathbf{B}) \wedge (\mathbf{B} \rightarrow \mathbf{A})$.

D5: $\top = -\bot$.

---

[12] The present work—as was explained in the preface—constitutes only a fragment of what was conceived as a much more comprehensive monograph. In fact, the Lemmon Notes, less the historical introduction, are the draft of the first chapter only.

The five belong to the usual propositional calculus. The next batch are more specifically modal:

$D6$: $\quad\Diamond \mathbf{A} = {-}\Box{-}\mathbf{A}$.
$D7(n)$: $\quad\Box^n \mathbf{A} = \underbrace{\Box \cdots \Box}_{n}\mathbf{A}$ (whence $\Box^0 \mathbf{A} = \mathbf{A}$).

$D8(n)$: $\quad\Diamond^n \mathbf{A} = \underbrace{\Diamond \cdots \Diamond}_{n}\mathbf{A}$ (whence $\Diamond^0 \mathbf{A} = \mathbf{A}$).

$D9(n)$: $\quad\mathbf{A} \Rightarrow^n \mathbf{B} = \Box^n(\mathbf{A} \to \mathbf{B})$
$\quad\quad\quad$ (whence $\Box^0(\mathbf{A} \to \mathbf{B}) = (\mathbf{A} \to \mathbf{B})$).
$D10$: $\quad\mathbf{A} \Rightarrow \mathbf{B} = \mathbf{A} \Rightarrow^1 \mathbf{B}$ (i.e., $\mathbf{A} \Rightarrow \mathbf{B}$ is $\Box(\mathbf{A} \to \mathbf{B})$).
$D11(n)$: $\quad\mathbf{A} \Leftrightarrow^n \mathbf{B} = (\mathbf{A} \Rightarrow^n \mathbf{B}) \wedge (\mathbf{B} \Rightarrow^n \mathbf{A})$.
$D12$: $\quad\mathbf{A} \Leftrightarrow \mathbf{B} = \mathbf{A} \Leftrightarrow^1 \mathbf{B}$ (i.e., $\mathbf{A} \Leftrightarrow \mathbf{B}$ is $(\mathbf{A} \Rightarrow \mathbf{B}) \wedge (\mathbf{B} \Rightarrow \mathbf{A})$).

Here $\Diamond$ represents possibility (compare (21) of the Introduction), $\Rightarrow$ strict implication (compare the Introduction), and $\Leftrightarrow$ the corresponding strict equivalence.

When convenient, in exhibiting sentences or schemes of sentences we shall drop brackets in conformity with normal conventions. Thus we shall not usually bother with the encircling brackets in $(\mathbf{A} \to \mathbf{B})$, writing $\mathbf{A} \to \mathbf{B}$. And in place of $(\mathbf{A} \wedge \mathbf{B}) \to \mathbf{C}$, $\mathbf{A} \to (\mathbf{B} \vee \mathbf{C})$, etc., we write $\mathbf{A} \wedge \mathbf{B} \to \mathbf{C}$, $\mathbf{A} \to \mathbf{B} \vee \mathbf{C}$, etc., though we shall write $\mathbf{A} \wedge (\mathbf{B} \to \mathbf{C})$, $(\mathbf{A} \to \mathbf{B}) \vee \mathbf{C}$, etc., to avoid ambiguity.

We study modal sentences in two quite different ways, a semantical and a syntactical. It is the first business of a semantical approach to state the conditions under which a modal sentence is true or false for a given interpretation; to this task we turn to Section 1 of this chapter. The syntactical approach, on the other hand, is more concerned with *systems* of modal sentences, in a sense we shall explain in the remainder of this section. A fruitful blend of the two approaches issues in a completeness result for a system: we obtain our first results of this kind in Section 2.

In the broadest sense, a *system* is a class of modal sentences, say Γ, with the property that if $\mathbf{A} \in \Gamma$, $\mathbf{A} \to \mathbf{B} \in \Gamma$, then $\mathbf{B} \in \Gamma$ ($\in$ represents class membership). (It is thus sometimes said that systems are *closed with respect to modus ponens*.) If S is any system, we refer to the members of S (sentences in S) as *theorems of S*, and usually write $\vdash_S \mathbf{A}$ in place of $\mathbf{A} \in S$. We shall regularly use the term 'system' in a rather narrower sense, however, whereby a system is understood to *include the classical propositional calculus* (compare, for example, the Gödel version of *S4* mentioned in the Introduction). Let us explain what this means.

Consider the schemata:

$A1(1)$: $\mathbf{A} \to (\mathbf{B} \to \mathbf{A})$.

$A1(2)$: $(\mathbf{A} \to (\mathbf{B} \to \mathbf{C})) \to ((\mathbf{A} \to \mathbf{B}) \to (\mathbf{A} \to \mathbf{C}))$.

$A1(3)$: $((\mathbf{A} \to \bot) \to \bot) \to \mathbf{A}$.

Any (modal) sentence having the form of either $A1(1)$, $A1(2)$, or $A1(3)$ is called an *axiom of propositional calculus*. Thus $\mathbf{P}_3 \to ((\mathbf{P}_2 \to \bot) \to \mathbf{P}_3)$, $((\square(\mathbf{P}_1 \wedge \mathbf{P}_6) \to \bot) \to \bot) \to \square(\mathbf{P}_1 \wedge \mathbf{P}_6)$ are propositional calculus axioms. We also use $A1$ to designate the *class* of such axioms, i.e., the class of sentences having one of the forms $A1(1)$–$A1(3)$. Then the *propositional calculus* (*PC*) is the smallest class Γ of sentences such that all members of $A1$ are in Γ ($A1 \subseteq \Gamma$) and such that, if $\mathbf{A} \in \Gamma$, $\mathbf{A} \to \mathbf{B} \in \Gamma$, then $\mathbf{B} \in \Gamma$.

Perhaps a more natural, certainly a more familiar, way of conceiving the propositional calculus is as follows. In addition to thinking of the members of $A1$ as *axioms*, we introduce a *rule of inference*:

MP: from $\mathbf{A}$ and $\mathbf{A} \to \mathbf{B}$, to derive $\mathbf{B}$.

A *proof* is then defined as a finite sequence of sentences such that each member either belongs to $A1$ or is derived from earlier members of the sequence by *MP*. A proof is said to be a proof *of* the last member in its sequence, and a *theorem* is a sentence of which there is a proof. We then define *PC* as the class of sentences which are theorems. But these two definitions of *PC* are equivalent (as the reader unfamiliar with these alternatives should verify). In practice, we shall usually specify systems by specifying their axioms and rules of inference; it is then to be understood that the system is the class of theorems generated from the axioms by the rules of inference in the sense just indicated. For such systems S, we may understand $\vdash_S \mathbf{A}$ to imply the existence of a finite sequence of sentences constituting a *proof* of $\mathbf{A}$ in S.

So by a system (in the narrower sense) we mean a system (in the broad sense) including $A1$. Thus a *system* from now on is a set of sentences Γ such that $A1 \subseteq \Gamma$ and if $\mathbf{A}, \mathbf{A} \to \mathbf{B} \in \Gamma$, then $\mathbf{B} \in \Gamma$. It clearly follows that $PC \subseteq S$, for any system S. In this sense a system includes the usual propositional calculus. It is known (see, for example, Church [2]) that $A1(1)$–$A1(3)$ provide a sufficient axiomatic basis for the derivation as theorems of all tautologies. Since we are assuming here familiarity with elementary logic, we small make use of the fact that any system contains any tautology without more ado. (We make no study of modal systems based on propositional systems other than the classical, such as intuitionistic logic.)

We proceed to define various properties of systems in the sense given. For *any* class of sentences Γ and system S, we shall say that $\mathbf{A}$ is *S-deducible* from Γ iff there are sentences $\mathbf{B}_1, \ldots, \mathbf{B}_n \in \Gamma$ such that $\vdash_S \mathbf{B}_1 \wedge \cdots \wedge \mathbf{B}_n \to \mathbf{A}$: in symbols $\Gamma \vdash_S \mathbf{A}$. (Here and hereafter 'iff' is short for 'if and only if', and the notation '$\mathbf{B}_1, \ldots, \mathbf{B}_n$' is meant to suggest some *finite* number n of sentences that may be indexed $1, \ldots, n$.) Given that $PC \subseteq S$ for any system S, it is easily shown that $\Gamma \vdash_S \mathbf{A}$ iff there is a finite sequence $\mathbf{A}_1, \ldots, \mathbf{A}_m$ of sentences such that $\mathbf{A}_m = \mathbf{A}$ and for each i ($1 \leq i \leq m$) either (i) $\mathbf{A}_i \in \Gamma$ or (ii) $\vdash_S \mathbf{A}_i$ or (iii) there are $j, k < i$ such that $\mathbf{A}_i$ is derivable from $\mathbf{A}_j$ and $\mathbf{A}_k$ by *MP*: this sequence $\mathbf{A}_1, \ldots, \mathbf{A}_m = \mathbf{A}$ may be called a *derivation of $\mathbf{A}$ from Γ in S*. This notion of S-deducibility is, of course, standard. In particular, the so-called *deduction theorem* holds: if $\Gamma \cup \{\mathbf{A}\} \vdash_S \mathbf{B}$ then $\Gamma \vdash_S \mathbf{A} \to \mathbf{B}$. It is also easy to see that the class of sentences $\mathbf{A}$ such that $\Gamma \vdash_S \mathbf{A}$ itself forms a system. Γ may further be the null class $\emptyset$, in which case $\Gamma \vdash_S \mathbf{A}$ iff $\vdash_S \mathbf{A}$.

If S is a system, by an *S-system* we mean a system S' including S, i.e., such that $S \subseteq S'$. S-systems are sometimes called *extensions* of S. S is itself an S-system, since $S \subseteq S$. More generally, for *any* class of sentences Γ, by an *S-extension of Γ* we mean an S-system including Γ, i.e., a system S' such that $\Gamma \subseteq S', S \subseteq S'$. Clearly, if $\Gamma \vdash_S \mathbf{A}$, then $\mathbf{A}$ belongs to all S-extensions of Γ. (For, given $\Gamma \vdash_S \mathbf{A}$, we have $\vdash_S \mathbf{B}_1 \wedge \cdots \wedge \mathbf{B}_n \to \mathbf{A}$ for $\mathbf{B}_1, \ldots, \mathbf{B}_n \in \Gamma$: select now S' an S-extension of Γ; then $\vdash_{S'} \mathbf{B}_1 \wedge \cdots \wedge \mathbf{B}_n \to \mathbf{A}$ since $S \subseteq S'$, and $\vdash_{S'} \mathbf{B}_1, \ldots, \vdash_{S'} \mathbf{B}_n$, since $\Gamma \subseteq S'$; it follows that $\vdash_{S'} \mathbf{A}$ since $PC \subseteq S'$.) The class of $\mathbf{A}$ such that $\Gamma \vdash_S \mathbf{A}$, or in set-theoretic notation $\{\mathbf{A} : \Gamma \vdash_S \mathbf{A}\}$, is itself an S-extension of Γ, and contained in all S-extensions of Γ.

A system S is *consistent* iff not all sentences belong to S; equivalently in the light of PC, iff $\bot \notin S$, i.e., not $\bot \in S$. Otherwise S is *inconsistent*. (There is only one inconsistent system; it is the class of all sentences.) A system S is *complete* iff for all sentences $\mathbf{A}$ either $\mathbf{A} \in S$ or $-\mathbf{A} \in S$ ($\mathbf{A} \to \bot \in S$). (The inconsistent system is trivially complete.)

More generally, a set of sentences Γ is *S-consistent* (consistent with respect to the system S) iff it is not the case that $\Gamma \vdash_S \mathbf{A}$ for all sentences **A**; equivalently, in the light of *PC*, iff not $\Gamma \vdash_S \bot$. Then $\emptyset$ (the null class) is *S*-consistent iff *S* is consistent. Also a set Γ of sentences is *S*-complete (complete with respect to *S*) iff for all sentences **A** either $\Gamma \vdash_S \mathbf{A}$ or $\Gamma \vdash_S -\mathbf{A}$. Then $\emptyset$ is *S*-complete iff *S* is complete. A system which is both consistent and complete is called a *maximal consistent* system; we shall have a great deal to do with such systems in the sequel.

We note some simple facts about systems *S* and sets of sentences Γ, which will be used extensively (and often tacitly) later:

(i) for any system S, $\mathbf{A} \wedge \mathbf{B} \in S$ iff $\mathbf{A} \in S$ and $\mathbf{B} \in S$;
(ii) for any consistent S, if $\mathbf{A} \in S$ then $-\mathbf{A} \notin S$ and if $-\mathbf{A} \in S$ then $\mathbf{A} \notin S$;
(iii) for any complete S, $\mathbf{A} \vee \mathbf{B} \in S$ iff $\mathbf{A} \in S$ or $\mathbf{B} \in S$;
(iv) for any complete S, $\mathbf{A} \to \mathbf{B} \in S$ iff if $\mathbf{A} \in S$ then $\mathbf{B} \in S$;
(v) for any maximal consistent S, $\mathbf{A} \in S$ iff $-\mathbf{A} \notin S$ and $-\mathbf{A} \in S$ iff $\mathbf{A} \notin S$;
(vi) if Γ is *S*-consistent, *S* is consistent;
(vii) if Γ is S'-consistent and S' is an *S*-system, then Γ is *S*-consistent.

For purposes of illustration, we demonstrate (iii). Suppose *S* is complete. If either $\mathbf{A} \in S$ or $\mathbf{B} \in S$, then $\mathbf{A} \vee \mathbf{B} \in S$ by *PC* (both $\mathbf{A} \to \mathbf{A} \vee \mathbf{B}$ and $\mathbf{B} \to \mathbf{A} \vee \mathbf{B}$ are tautologies.) Conversely, suppose neither $\mathbf{A} \in S$ nor $\mathbf{B} \in S$; then clearly *S* is consistent, since otherwise $\mathbf{A} \in S$ for all **A**; by the completeness of S, $-\mathbf{A} \in S$ and $-\mathbf{B} \in S$ so that $-\mathbf{A} \wedge -\mathbf{B} \in S$, i.e., $-(\mathbf{A} \vee \mathbf{B}) \in S$, by *PC*; thus $\mathbf{A} \vee \mathbf{B} \notin S$ by (ii) and the consistency of *S*. This gives (iii).

We now prove two basic facts about systems and sets of sentences, both of which will play a considerable role in our later completeness results.

Theorem 0.1. (Lindenbaum's Lemma). *Every S-consistent set Γ of sentences has a maximal consistent S-extension.*

Proof. Let Γ be an *S*-consistent set of sentences, and consider some enumeration $\mathbf{A}_1, \mathbf{A}_2, \ldots, \mathbf{A}_n, \ldots$ of all modal sentences. We define a series of *S*-extensions of Γ as follows: S_0 is the set of all **A** such that $\Gamma \vdash_S \mathbf{A}$ (thus S_0 is consistent as well as being an *S*-extension of Γ); assume S_n defined; then if $S_n \vdash_S -\mathbf{A}_{n+1}$, put $S_{n+1} = S_n$, but if not $S_n \vdash_S -\mathbf{A}_{n+1}$, put S_{n+1} the set of all **A** such that $S_n \cup \{\mathbf{A}_{n+1}\} \vdash_S \mathbf{A}$. We show that S_n is a consistent *S*-extension of Γ for any n. This holds for $n = 0$, as we have seen. Assume that it holds for S_n. If $S_{n+1} = S_n$, S_{n+1} is of course a consistent *S*-extension of Γ, so assume

$S_{n+1} \neq S_n$. Then not $S_n \vdash_S -\mathbf{A}_{n+1}$ by definition of S_{n+1}. However, if S_{n+1} were inconsistent, we would have $S_n \cup \{\mathbf{A}_{n+1}\} \vdash_S \bot$, so that $S_n \vdash_S \mathbf{A}_{n+1} \to \bot$ by the deduction theorem, which is a contradiction. So S_{n+1} is a consistent S-extension of Γ. Now put $S' = \bigcup \{S_n : n = 0, 1, \ldots\}$, i.e., S' is the set of all sentences occurring in some S_n. It is easy to verify that S' is indeed a system (closed with respect to MP), and an S-extension of Γ. Further, S' is consistent, for if $\vdash_{S'} \bot$ then $\vdash_{S_n} \bot$ for some n, contradicting the consistency of all S_n. To show that S' is complete, let $\mathbf{A}$ be any modal sentence, say $\mathbf{A} = \mathbf{A}_{n+1}$ for $n \geq 0$. By definition of S_{n+1}, we have either $S_n \vdash_S -\mathbf{A}_{n+1}$ or $S_{n+1} \vdash_S \mathbf{A}_{n+1}$, so that either $\vdash_{S'} \mathbf{A}$ or $\vdash_{S'} -\mathbf{A}$. Thus S' is a maximal consistent S-extension of Γ.

Theorem 0.2. *For any set of sentences* Γ, $\Gamma \vdash_S \mathbf{A}$ *iff* $\mathbf{A}$ *belongs to all maximal consistent S-extensions of Γ.*

Proof. If $\Gamma \vdash_S \mathbf{A}$, then $\mathbf{A}$ belongs to *all* S-extensions of Γ, as remarked above. Conversely, suppose not $\Gamma \vdash_S \mathbf{A}$, and consider $\Gamma' = \Gamma \cup \{-\mathbf{A}\}$. Then Γ' is S-consistent, since otherwise $\Gamma' \vdash_S \bot$, whence $\Gamma \vdash_S -\mathbf{A} \to \bot$ by the deduction theorem and $\Gamma \vdash_S \mathbf{A}$ by PC. Hence Γ' has a maximal consistent S-extension S' by 0.1, which is also a maximal consistent S-extension of Γ since $\Gamma \subseteq \Gamma'$. But $-\mathbf{A} \in S'$, so that $\mathbf{A} \notin S'$ by the consistency of S'. This completes the proof.

Corollary. $\vdash_S \mathbf{A}$ *iff* $\mathbf{A}$ *belongs to all maximal consistent S-systems.*

Proof. Take Γ as $\emptyset$ in the theorem.

So far our discussion has been quite general, in that no specific attention has been paid to the presence or absence of $\Box$ in modal sentences belonging to systems. Reverting to our discussion in the Introduction, we say that a system S is *classical* iff it provides the rule RE, i.e., iff whenever $\vdash_S \mathbf{A} \leftrightarrow \mathbf{B}$ then $\vdash_S \Box \mathbf{A} \leftrightarrow \Box \mathbf{B}$. We remind the reader that, as a rule of derivation, RE has the form:

$$RE: \quad \text{from } \mathbf{A} \leftrightarrow \mathbf{B}, \text{ to derive } \Box \mathbf{A} \leftrightarrow \Box \mathbf{B}.$$

We now show:

Theorem 0.3. *If S is classical, then if* $\vdash_S \mathbf{A} \leftrightarrow \mathbf{B}$ *then* $\vdash_S \cdots \mathbf{A} \cdots \leftrightarrow \cdots \mathbf{B} \cdots$ (*where* $\cdots \mathbf{B} \cdots$ *results from* $\cdots \mathbf{A} \cdots$ *by replacing* $n (n \geq 0)$ *occurrences of* $\mathbf{A}$ *in* $\cdots \mathbf{A} \cdots$ *by* $\mathbf{B}$).

Proof is by induction on the complexity of the context $\cdots \mathbf{A} \cdots$ of $\mathbf{A}$. Where $\cdots \mathbf{A} \cdots$ is just $\mathbf{A}$, the result is obvious. For the inductive steps, we note first that by $PC \vdash_S (\mathbf{A} \leftrightarrow \mathbf{B}) \to ((\mathbf{A} \to \mathbf{C}) \leftrightarrow (\mathbf{B} \to \mathbf{C}))$

and $\vdash_S (\mathbf{A} \leftrightarrow \mathbf{B}) \to ((\mathbf{C} \to \mathbf{A}) \leftrightarrow (\mathbf{C} \to \mathbf{B}))$. The other case is covered by *RE* itself, given that S is classical.

In virtue of 0.3, any classical system S provides the full substitutivity of *provable* material equivalents, **A** and **B** such that $\vdash_S \mathbf{A} \leftrightarrow \mathbf{B}$. Another rule mentioned in the Introduction, namely

RM: from $\mathbf{A} \to \mathbf{B}$, to derive $\Box \mathbf{A} \to \Box \mathbf{B}$,

is such that, if a system S provides it, then it provides *RE*. For suppose S provides *RM*, i.e., if $\vdash \mathbf{A} \to \mathbf{B}$ then $\vdash_S \Box \mathbf{A} \to \Box \mathbf{B}$, and suppose $\vdash_S \mathbf{A} \leftrightarrow \mathbf{B}$. Then $\vdash_S \mathbf{A} \to \mathbf{B}$, $\vdash_S \mathbf{B} \to \mathbf{A}$ by *PC*, whence $\vdash_S \Box \mathbf{A} \to \Box \mathbf{B}$, $\vdash_S \Box \mathbf{B} \to \Box \mathbf{A}$ by *RM*. This gives $\vdash_S \Box \mathbf{A} \leftrightarrow \Box \mathbf{B}$ by *PC*, demonstrating *RE*. Thus any system with *RM* is classical, and provides the substitutivity of provable material equivalents. The converse of this is not true: we shall later find systems which provide *RE* but do not provide the rule *RM*.[13]

We may informally think of $\vdash_S \mathbf{A} \leftrightarrow \mathbf{B}$ as meaning that **A** and **B** are *intensionally equivalent* in S. If S is classical, 0.3 provides for the substitutivity in all contexts (including modal ones) of intensional equivalents.

---

[13] The reader will find no such systems in this work, as they were to be dealt with in a later chapter (cf. n. 12). However, using the so-called neighbourhood semantics—a generalization of Kripke-type semantics—it is easy to find systems of the kind Lemmon mentions. The smallest classical system is a case in point.

Section 1

It is well known that Leibniz suggested that necessity was equivalent to truth in all possible worlds. As a *definition* of necessary truth, however, this will hardly do as it stands, since the words 'possible' and 'world' remain unexplained. Of these two notions, that of 'world' perhaps presents the less difficulty; it might be explicated along the lines of Carnap's state descriptions (see Carnap [2]) or, for certain purposes, simply taken as primitive.[14] For the present, we shall assume that we understand well enough what is meant by a world, in such a way that it makes sense to speak of worlds alternative to the actual one. So by a *possible* world we may simply mean one which is an alternative to this one. Leibniz's suggestion now becomes: a sentence is necessarily true (in this world) iff that sentence is true in all worlds alternative to this world.

It may well be objected at this point that, although perhaps the notion of alternative worlds can be *fairly* well understood, the notion of a world alternative *to this one* is wretchedly obscure. Probably the only reply at this stage is that it is *meant* to be. For there are no doubt many kinds of necessity (or, if you prefer, many different senses which we hope to attach, in different contexts, to the primitive symbol □); and our hope, semantically speaking, is to clarify as many as possible by making precise *in different ways* the notion of alternativeness. However, some illustrations may help. A world t may be called a *scientific* alternative to a world u if at least all scientific laws which hold in u also hold in t. Or a world t may be said to be a *moral* alternative to a world u if at least all moral laws which hold in u continue to hold in t. Or a world t may be called a *logical* alternative to a world u if at least all logical laws holding in u hold in t. Thus, if in a world t a body travels at 200,000 miles per second, t may be a logical alternative to the actual world, but is not a scientific alternative to it. And, if in t an act of wanton murder is morally correct, then t may be a scientific alternative to the actual

[14] Considering its importance for applied modal logic it is perhaps astonishing that the notion of possible world has not received more attention in the literature than is actually the case. However, some interesting references have appeared since Lemmon's time. Among them are M. J. Cresswell, *Logics and languages* (London: Methuen & Co., 1973), David Lewis, *Counterfactuals* (Oxford: Basil Blackwell, 1973), Alvin Plantinga, *The nature of necessity* (Oxford: At the Clarendon Press, 1974) and Dana Scott, 'Advice on modal logic', *Philosophical problems in logic*, edited by K. Lambert, pp. 143–73 (Dordrecht: Reidel, 1970).

world, but is not a moral alternative to it. If we explain 'scientifically (morally, logically) necessary' as meaning 'true in all scientific (moral, logical) alternative worlds', we hope that the distinctions between these notions—such as they are—may emerge by consideration of the different properties of alternativeness employed in the explanations. Hence it is all to the good at the outset to lay down *no* conditions which the alternativeness relation *must* satisfy.

Actually, in many connections it is intuitively simpler to think of world t as *accessible from* world u rather than *alternative to u*. This at least has the merit of avoiding the temptation to suppose that alternativeness is a symmetric relation between worlds—that if t is alternative to u, then u must be alternative to t. In pursuance of our policy of making the fewest possible assumptions, we want to leave open the possibility that t may be accessible from u though u not accessible from t. Indeed, we shall not assume that each world is accessible from itself, or even that to each world there is at least one accessible world: there may be accessibility-isolated worlds. We shall find that to many such assumptions about the accessibility relation between worlds there correspond distinctive modal sentences which come out valid precisely because we have made those assumptions.

If necessity means truth in all accessible worlds, then possibility will mean truth in some accessible world. Thus our remarks about the vagueness of the notion of necessity, and the various more precise accounts of it, may be repeated *mutatis mutandis* for the notion of possibility.

More formally, then, we introduce the notion of a system of worlds, or *world system* (w.s.). A world system is to consist of a set of worlds and an accessibility relation defined on that set; thus let U be any non-empty set of elements, to be called worlds; and R any relation on U, to be called the accessibility relation for U; then $\mathcal{K} = \langle U, R \rangle$ is a world system. If we think of U as a universe, then a world system not only fixes what worlds there are in the universe but which is accessible from which. Note that *no* conditions are imposed on the relation R, not even that it be non-empty.

If we are to employ world systems $\mathcal{K}$ for the purpose of giving suitable semantic definitions of truth, validity, and the like, for the sentences of an intensional logic, we need in the first place to state what the truth value of each atomic sentence is to be at each world $u \in U$. At the propositional calculus level, this involves specifying, for each sentence letter $\mathbf{P}_i$, a subset $P_i \subseteq U$ as the set of worlds at which $\mathbf{P}_i$ is true—it being understood that at each $u \notin P_i$ $\mathbf{P}_i$ is false.

Thus we define a *modal structure* (m.s.) as a structure

$$\mathcal{U} = \langle U, R, P_0 \ldots, P_n, \ldots \rangle,$$

where U is again a non-empty set and R a relation on U, and further each $P_i \subseteq U$. Corresponding to a given w.s. $\mathcal{K}$, there will evidently be (at least denumerably) many different modal structures, depending on the choice of the P_i. We shall call these modal structures *on* $\mathcal{K}$. Conversely, given an m.s. $\mathcal{U} = \langle U, R, P_0, \ldots, P_n, \ldots \rangle$, we shall say that $\mathcal{K} = \langle U, R \rangle$ is the *world system corresponding to* $\mathcal{U}$.

It is often convenient to describe a modal structure as of the form $\mathcal{U} = \langle U, R, \varphi \rangle$, where φ is simply a function from the natural numbers to subsets of U. Given such a description, we can recreate the original modal structure as simply

$$\langle U, R, \varphi(0), \ldots, \varphi(n), \ldots \rangle.$$

From this standpoint, the function φ determines for each sentence letter $\mathbf{P}_i$ a set $\varphi(i)$ of worlds at which $\mathbf{P}_i$ is true; equivalently, for each world $u \in U$ φ determines a set of sentence letters true at u—namely, the set $\{\mathbf{P}_i : u \in \varphi(i)\}$. Compare Kripke [1] and [2].

At last we can define what it is for a sentence $\mathbf{A}$ to be *true at* a world u *in* a modal structure $\mathcal{U} = \langle U, R, P_0, \ldots, P_n, \ldots \rangle (u \in U)$. We abbreviate all this to

$$\vDash_u^{\mathcal{U}} \mathbf{A}.$$

The definition is of course recursive:

(i) if $\mathbf{A}$ is a sentence letter $\mathbf{P}_i$, then

$$\vDash_u^{\mathcal{U}} \mathbf{A} \quad \text{iff } u \in P_i;$$

(ii) if $\mathbf{A}$ is $\bot$, then

$$\vDash_u^{\mathcal{U}} \mathbf{A} \quad \text{iff the false (i.e., not} \quad \vDash_u^{\mathcal{U}} \mathbf{A});$$

(iii) if $\mathbf{A}$ has the form $\mathbf{B} \to \mathbf{C}$, then

$$\vDash_u^{\mathcal{U}} \mathbf{A} \quad \text{iff if} \quad \vDash_u^{\mathcal{U}} \mathbf{B} \text{ then } \quad \vDash_u^{\mathcal{U}} \mathbf{C};$$

(iv) if $\mathbf{A}$ has the form $\Box \mathbf{B}$, then

$$\vDash_u^{\mathcal{U}} \mathbf{A} \quad \text{iff for all } t \text{ such that} \quad uRt \quad \vDash_t^{\mathcal{U}} \mathbf{B}.$$

This definition will be referred to as the (first) *truth definition* (there are others to follow).[15] Clause (i) corresponds to our informal explanation of the role of the P_i above. If $\mathcal{U}$ is specified in terms of φ, it reads of course:

(i) if **A** is a sentence letter $\mathbf{P}_i$, then

$$\models_u^{\mathcal{U}} \mathbf{A} \quad \text{iff } u \in \varphi(i).$$

Clauses (ii) and (iii) are simply repeats of the usual propositional calculus truth clauses. In conjunction with our definitions of $-$, $\wedge$, and $\vee$ in the last section, they yield the familiar truth tables for $\rightarrow$, $-$, etc. Clause (iv), which requires that $\Box \mathbf{B}$ is true at u in $\mathcal{U}$ iff **B** is true at all t accessible from u, conforms to our intuitive account of the semantics of necessity earlier in this section. It is easy to verify that, as a result of this definition together with our definition of $\Diamond \mathbf{A}$ as $-\Box - \mathbf{A}$, we have,

(v) if **A** has form $\Diamond \mathbf{B}$, then

$$\models_u^{\mathcal{U}} \mathbf{A} \quad \text{iff there exists } t \text{ such that } uRt \text{ and } \models_t^{\mathcal{U}} \mathbf{B}.$$

So $\Diamond \mathbf{B}$ is true at u in $\mathcal{U}$ iff there is a world t accessible from u at which **B** is true, in conformity with our earlier discussion.

It will be worth our while at this point also to establish the truth conditions for $\Box^n \mathbf{A}$ and $\Diamond^n \mathbf{A}$ (for arbitrary n) in terms of those for **A**. To this end, we define R^n to be the relative product of R with itself n times. More precisely, we put:

$$uR^0 t \quad \text{iff} \quad u = t,$$
$$uR^{n+1} t \quad \text{iff} \quad (\exists t')(uRt' \wedge t'R^n t),$$

for any relation R. (We use propositional calculus and predicate calculus notation here and hereafter informally in our *metalanguage*, to shorten—and clarify—our statements.) Then we find,

(vi) $\models_u^{\mathcal{U}} \Box^n \mathbf{A}$ iff $(\forall t)(uR^n t \rightarrow \models_t^{\mathcal{U}} \mathbf{A})$;
(vii) $\models_u^{\mathcal{U}} \Diamond^n \mathbf{A}$ iff $(\exists t)(uR^n t \wedge \models_t^{\mathcal{U}} \mathbf{A})$.

(vi) may be proved by induction on n. The case $n = 0$ is just that $\models_u^{\mathcal{U}} \mathbf{A}$ iff $(\forall t)(u = t \rightarrow \models_t^{\mathcal{U}} \mathbf{A})$, which is logically true. Suppose the result holds for n. Then $\models_u^{\mathcal{U}} \Box^{n+1} \mathbf{A}$ iff $(\forall t)(uRt \rightarrow \models_t^{\mathcal{U}} \Box^n \mathbf{A})$ (by

---

[15] No other truth definitions are given in this work (unless the δ-semantics on p. 60 f. is counted as one). Presumably Lemmon had in mind semantics for tense logic and for non-normal systems; cf. n. 12.

(iv)) iff $(\forall t)(uRt \to (\forall t')(tR^n t' \to \vDash_{t'}^{\mathcal{U}} \mathbf{A}))$ (by the inductive hypothesis) iff $(\forall t')((\exists t)(uRt \wedge tR^n t') \to \vDash_{t'}^{\mathcal{U}} \mathbf{A})$ (by quantifier reasoning) iff $(\forall t)(uR^{n+1} t \to \vDash_{t}^{\mathcal{U}} \mathbf{A})$ (by definition of R^{n+1}). (vii) follows at once from (vi).

Intuitively, (vi) says that $\square^n \mathbf{A}$ is true at u in $\mathcal{U}$ iff for all worlds 'n-accessible' from u $\mathbf{A}$ is true in $\mathcal{U}$, where t is n-accessible from u if t can be reached from u in n steps of the accessibility relation R.

For a given world system $\mathcal{K} = \langle U, R \rangle$, we say that $\mathbf{A}$ is *valid in* $\mathcal{K}$—in symbols $\vDash^{\mathcal{K}} \mathbf{A}$—iff $\vDash_{u}^{\mathcal{U}} \mathbf{A}$ for all $\mathcal{U}$ on $\mathcal{K}$ and $u \in U$; and $\mathbf{A}$ is *satisfiable in* $\mathcal{K}$ iff $\vDash_{u}^{\mathcal{U}} \mathbf{A}$ for some $\mathcal{U}$ on $\mathcal{K}$ and $u \in U$. Finally, $\mathbf{A}$ is (simply) *valid* iff $\mathbf{A}$ is valid in all $\mathcal{K}$, and $\mathbf{A}$ is (simply) *satisfiable* iff $\mathbf{A}$ is satisfiable in some $\mathcal{K}$. That $\mathbf{A}$ is valid we write

$$\vDash \mathbf{A}.$$

Note that we might equivalently have said: $\mathbf{A}$ is valid iff $\vDash_{u}^{\mathcal{U}} \mathbf{A}$ for all $\mathcal{U} = \langle U, R, \varphi \rangle$ and $u \in U$; $\mathbf{A}$ is satisfiable iff $\vDash_{u}^{\mathcal{U}} \mathbf{A}$ for some $\mathcal{U} = \langle U, R, \varphi \rangle$ and $u \in U$.

In order to clarify further this notion of validity, it may be as well to compare it with that for the usual propositional calculus: namely, tautologousness (by truth table). Since in this case we are not concerned with intensional concepts like necessity, we assume that there is exactly one world, the actual one, say u, and we may as well assume that it is accessible from itself, uRu. Accordingly, our world systems $\mathcal{K} = \langle U, R \rangle$ all have the form $\langle \{u\}, \{\langle u, u \rangle\} \rangle$ for some u. Any $\mathcal{U}$ on such a $\mathcal{K}$ has only two possibilities for its P_i, namely $\{u\}$ and $\emptyset$. Thus such a $\mathcal{U}$ is really just a simultaneous assignment of the truth values $\top$ and $\bot$ to all $\mathbf{P}_i$: if P_i is $\{u\}$, $\mathbf{P}_i = \top$, and if $P_i = \emptyset$, $\mathbf{P}_i = \bot$. In this situation clause (iv) of the truth definition collapses necessary truth into truth: $\vDash_{u}^{\mathcal{U}} \square \mathbf{A}$ iff $\vDash_{u}^{\mathcal{U}} \mathbf{A}$. Validity of $\mathbf{A}$ means truth for all possible such simultaneous assignments, and this can easily be shown equivalent to tautologousness under a truth table test.

In addition to the notion of validity of $\mathbf{A}$ in a *world system* $\mathcal{K}$, just defined, it is often convenient to employ the notion of $\mathbf{A}$'s validity in a *modal structure* $\mathcal{U} = \langle U, R, \varphi \rangle$, which we write $\vDash^{\mathcal{U}} \mathbf{A}$ (hoping that the superscript $\mathcal{U}$ rather than $\mathcal{K}$ will distinguish the two ideas sufficiently). We say that $\mathbf{A}$ is *valid* in $\mathcal{U}$ iff $\vDash_{u}^{\mathcal{U}} \mathbf{A}$ for all $u \in U$. Obviously, $\vDash^{\mathcal{K}} \mathbf{A}$ iff $\vDash^{\mathcal{U}} \mathbf{A}$ for all $\mathcal{U}$ on $\mathcal{K}$, and $\vDash \mathbf{A}$ iff $\vDash^{\mathcal{U}} \mathbf{A}$ for all $\mathcal{U}$.

The recent history of the semantics given here (in particular, of clause (iv) of the truth definition) is a little difficult to determine. There is no doubt that full credit for seeing its importance for modal logic must go to Kripke (see [1] and [2]), but there were several anticipations.[16] The notion of alternative worlds has been widely exploited by Hintikka (see [2]), as well as by Prior, who derived the idea from some unpublished notes of C. A. Meredith concerning 'the calculus of properties' (see Thomas [1]). That the relation R may be intuitively thought of as a relation between possible worlds is, according to Prior, due to Peter Geach.[17] See also Kaplan's review of Kripke [2] (Kaplan [1]).

We devote the remainder of this section to establishing certain rather general results concerning truth and world systems; they will play a recurrent role in what follows. In the first place, it turns out that no loss of generality occurs if we restrict attention to a certain subclass of world systems which we may call *connected*.[18] (This point is made in Kripke [2]).

For $\mathcal{K} = \langle U, R \rangle$, let R^* be the ancestral of R, i.e.,

$$uR^*t \text{ iff for all sets } A \subseteq U \text{ such that } u \in A \text{ and } R``A \subseteq A \; t \in A,$$

where

$$R``A = \{t : (\exists t')(t' \in A \wedge t'Rt)\}.$$

We say $\mathcal{K}$ is *connected* iff there exists $u \in U$ such that $U = \{u\} \cup \{t : uR^*t\}$. Intuitively, $\mathcal{K} = \langle U, R \rangle$ is connected if we can find a world $u \in U$ such that U consists of u together with u's 'descendants' by the relation R. Further, for $\mathcal{K} = \langle U, R \rangle$ and $u \in U$, by $\mathcal{K}_u$, *the connected world system generated by u from $\mathcal{K}$*, we understand the

---

[16] It seems that Stig Kanger's *Provability in logic* should have been mentioned here (see footnote 5 on p. 8). Jaakko Hintikka's early paper *Quantifiers in deontic logic*, Societas Scientiarum Fennica, Commentationes humanarum litterarum, vol. 23, no. 4 (Helsingfors, 1957) introduces the concept of a 'relation of copermission' (between model sets). Another work worth mentioning is B. Jónsson and A. Tarski, 'Boolean algebras with operators', *American journal of mathematics*, vol. 73 (1951), pp. 891–939. And in retrospect Prior [3] seems highly suggestive.

[17] In a letter to Dana Scott, Arthur Prior objected to this passage: 'What Geach contributed was not the interpretation of R as a relation between worlds (God knows when *that* started), but the interpretation of R as a relation of *accessibility*.' Cf. the first page of Prior's 'Possible worlds', *Philosophical quarterly*, vol. 12 (1962), pp. 36–43.

[18] This use of the word 'connected', which seems to originate with Kripke, cannot be recommended: in mathematics and logic the word is used quite differently. It would have been possible, and in the editor's opinion desirable, to delete 'connected' in expressions of the type 'connected world system generated by u' and replace it by 'generated' in all other contexts. However, as this issue is a matter of taste, Lemmon's text has not been altered.

w.s. $\langle U_u, R_u \rangle$, where

$$U_u = \{u\} \cup \{t : uR^*t\},$$
$$R_u = \{\langle t, t' \rangle : tRt' \wedge t, t' \in U_u\}.$$

(It is obvious that $\mathcal{K}_u$, as defined, *is* connected.) Finally, for $\mathcal{U}$ on $\mathcal{K}$, $\mathcal{U} = \langle U, R, \varphi \rangle$, we define $\mathcal{U}_u$ on $\mathcal{K}_u$, $\mathcal{U}_u = \langle U_u, R_u, \varphi_u \rangle$, where for each i

$$\varphi_u(i) = \varphi(i) \cap U_u.$$

Thus $\mathcal{K}_u$ is $\mathcal{K}$ restricted to u and its 'descendants', and $\mathcal{U}_u$ is $\mathcal{U}$ restricted to $\mathcal{U}_u$. We can now prove

Theorem 1.1. *For any* $\mathcal{U} = \langle U, R, \varphi \rangle$, $u \in U$, $t \in U_u$,

$$\vDash_t^{\mathcal{U}} \mathbf{A} \quad iff \quad \vDash_t^{\mathcal{U}_u} \mathbf{A}.$$

Proof by induction on the length of **A**. Many of our proofs have this form; for the record, we give this one fairly fully; in the future we shall be more sparing.

Fix $\mathcal{U} = \langle U, R, \varphi \rangle$, pick $u \in U$. We show that, for any $t \in U_u$, $\vDash_t^{\mathcal{U}} \mathbf{A}$ iff $\vDash_t^{\mathcal{U}_u} \mathbf{A}$. Consider first the case where **A** is an atomic sentence $\mathbf{P}_i$. Then

$$\vDash_t^{\mathcal{U}} \mathbf{A} \quad \text{iff} \quad t \in \varphi(i),$$
$$\vDash_t^{\mathcal{U}_u} \mathbf{A} \text{ iff } t \in \varphi_u(i).$$

But, on the assumption that $t \in U_u$, these right-hand sides are clearly equivalent, by definition of $\varphi_u(i)$. So the claim is correct here. Similarly, if **A** is $\bot$, then not $\vDash_t^{\mathcal{U}} \mathbf{A}$ and not $\vDash_t^{\mathcal{U}_u} \mathbf{A}$, so the claim is correct here too.

Suppose **A** has the form $\mathbf{B} \to \mathbf{C}$, and pick $t \in U_u$. Then

$$\vDash_t^{\mathcal{U}} \mathbf{A} \quad \text{iff if} \vDash_t^{\mathcal{U}} \mathbf{B} \text{ then} \quad \vDash_t^{\mathcal{U}} \mathbf{C};$$
$$\vDash_t^{\mathcal{U}_u} \mathbf{A} \quad \text{iff if} \vDash_t^{\mathcal{U}_u} \mathbf{B} \text{ then} \quad \vDash_t^{\mathcal{U}_u} \mathbf{C}.$$

The right-hand sides are here equivalent by the inductive hypothesis.

Suppose, finally, that **A** has the form $\Box \mathbf{B}$, and pick $t \in U_u$. Then

$$\vDash_t^{\mathcal{U}} \mathbf{A} \quad \text{iff} \quad (\forall t')(tRt' \to \vDash_{t'}^{\mathcal{U}} \mathbf{B}),$$
$$\vDash_t^{\mathcal{U}_u} \mathbf{A} \quad \text{iff} \quad (\forall t')(tR_u t' \to \vDash_{t'}^{\mathcal{U}_u} \mathbf{B}).$$

Let the right-hand side of the first biconditional be (1), and the right-hand side of the second biconditional be (2). We need to show

(1) iff (2). So suppose (1), and tR_ut' for arbitrary t'. By definition of R_u, tRt' and $t, t' \in U_u$. Since tRt', $\models_{t'}^{\mathcal{U}} \mathbf{B}$ by (1). Since $t' \in U_u$, $\models_{t'}^{\mathcal{U}_u} \mathbf{B}$ by the inductive hypothesis. This gives (2). So now suppose (2), and tRt' for arbitrary t'. Since $t \in U_u$ by hypothesis, it follows by the definition of ancestral that $t' \in U_u$. Hence tR_ut' by definition of R_u. Using (2), we conclude $\models_{t'}^{\mathcal{U}_u} \mathbf{B}$. Given $t' \in U_u$, by the inductive hypothesis it follows that $\models_{t'}^{\mathcal{U}} \mathbf{B}$. This gives (1), and completes the proof of Theorem 1.1.

Corollary 1. *For any* $\mathcal{U} = \langle U, R, \varphi \rangle$, $u \in U$, *if* $\models_u^{\mathcal{U}} \mathbf{A}$ *then* $\models_u^{\mathcal{U}_u} \mathbf{A}$.

Proof. Select $\mathcal{U} = \langle U, R, \varphi \rangle$, and suppose $\models_u^{\mathcal{U}} \mathbf{A}$. By definition $u \in U_u$. By the theorem, it follows that $\models_u^{\mathcal{U}_u} \mathbf{A}$.

Corollary 2. $\models \mathbf{A}$ *iff* $\models^{\mathcal{K}} \mathbf{A}$ *for all connected* $\mathcal{K}$.

Proof. If $\models \mathbf{A}$, then clearly $\models^{\mathcal{K}} \mathbf{A}$ for all connected $\mathcal{K}$. Conversely, suppose not $\models \mathbf{A}$. Then we have $\mathcal{U} = \langle U, R, \varphi \rangle$ and $u \in U$ such that not $\models_u^{\mathcal{U}} \mathbf{A}$, i.e., $\models_u^{\mathcal{U}} - \mathbf{A}$. Hence $\models_u^{\mathcal{U}_u} - \mathbf{A}$ by Corollary 1. But $\mathcal{K}_u$ is connected. Thus we find $\mathcal{K}$ such that not $\models^{\mathcal{K}} \mathbf{A}$.

This latter corollary gives the sense in which we may restrict attention to connected $\mathcal{K}$. The intuitive content of Theorem 1.1 is that, in establishing whether $\mathbf{A}$ is true at t in U, only worlds *later* than t in the R-relation—'descendants' of t—(as well as t itself) are relevant. This further suggests that, if we *add* to a world system $\mathcal{K}$ further worlds and extend R to a relation R' by adding only links from new worlds to old, then we shall not affect truth or falsity in any $\mathcal{U}$ on $\mathcal{K}$. This suggestion is made precise in the following theorem.

Put $\mathcal{K} = \langle U, R \rangle$, and let $\mathcal{K}' = \langle U', R' \rangle$ be any world system such that $U' = U \cup A$ for some A disjoint from U and $R' = R \cup S$ for some relation S such that if uSt then $u \in A$. Then we say $\mathcal{K}'$ is a *safe extension* of $\mathcal{K}$. Intuitively, $\mathcal{K}'$ is a safe extension of $\mathcal{K}$ if U' results from U by adding some 'new' worlds A and R' results from R by adding to R only 'leads' *from* A. More clearly, if $\mathcal{K}'$ is a safe extension of $\mathcal{K}$, then for $u \in U$, uRt iff $uR't$. (If uRt, then trivially $uR't$; if $uR't$, then either uRt or uSt; but if uSt, then $u \in A$, which contradicts the assumption that A is disjoint from U.) This fact makes the induction behind the next theorem straightforward.

Theorem 1.2. *If* $\mathcal{K}'$ *is a safe extension of* $\mathcal{K}$, *then for* $\mathcal{U} = \langle U, R, \varphi \rangle$ *on* $\mathcal{K}$ *and* $\mathcal{U}' = \langle U', R', \varphi \rangle$ *on* $\mathcal{K}'$, *and for any* $u \in U$

$$\models_u^{\mathcal{U}} \mathbf{A} \quad \text{iff} \quad \models_u^{\mathcal{U}'} \mathbf{A}.$$

Corollary. *If* $\mathcal{K}'$ *is a safe extension of* $\mathcal{K}$, *then if* $\models^{\mathcal{K}'} \mathbf{A}$ *then* $\models^{\mathcal{K}} \mathbf{A}$.

Proof. Suppose $\mathcal{K}'$ a safe extension of $\mathcal{K}$ and yet not $\models^{\mathcal{K}}\mathbf{A}$. Then we find $\mathcal{U}$ on $\mathcal{K}, u \in U$, such that $\models^{\mathcal{U}}_{u} - \mathbf{A}$. Hence by the theorem $\models^{\mathcal{U}'}_{u} - \mathbf{A}$ for $\mathcal{U}'$ on $\mathcal{K}'$. Thus not $\models^{\mathcal{K}'}\mathbf{A}$.

Of course, the converse of this corollary is not in general true—we may have $\models^{\mathcal{K}}\mathbf{A}$, yet not $\models^{\mathcal{K}'}\mathbf{A}$ for $\mathcal{K}'$ a safe extension of $\mathcal{K}$. Nonetheless, we shall on occasion be able to construct particular safe extensions of $\mathcal{K}$ which have properties in common with $\mathcal{K}$ such that the converse holds. When this happens, the corollary will tell us that validity in $\mathcal{K}$ is equivalent to validity in the particular $\mathcal{K}'$.

Section 2

We saw in the last section how a natural definition of validity for modal sentences can be given in terms of modal structures or world systems. We are now concerned to characterize *syntactically* the class of valid modal sentences. We shall in fact introduce a system K (named after Kripke) whose theorems are exactly the valid sentences. The proof of this result will constitute a completeness result for the system K. It also constitutes our first attempt to bring semantical and syntactical considerations together. In anticipation of completeness results for many other systems of modal logic in later sections, we give results here in as general a form as possible. As a consequence, our later results will be very easy consequences of the work in this section, which is therefore in a way the basis of the rest of the book. First, some preliminary results.

Theorem 2.1. *For any m.s.* $\mathcal{U}$ *if* $\models^{\mathcal{U}} \mathbf{A}$ *then* $\models^{\mathcal{U}} \Box \mathbf{A}$.

Proof. Suppose $\models^{\mathcal{U}} \mathbf{A}$ for $\mathcal{U} = \langle U, R, \varphi \rangle$, i.e., $\models_u^{\mathcal{U}} \mathbf{A}$ for all $u \in U$. Select any $u \in U$, and suppose uRt. Then by hypothesis $\models_t^{\mathcal{U}} \mathbf{A}$. Hence by the truth definition $\models_u^{\mathcal{U}} \Box \mathbf{A}$. Thus $\models^{\mathcal{U}} \Box \mathbf{A}$.

Corollary 1. *For any w.s.* $\mathcal{K}$, *if* $\models^{\mathcal{K}} \mathbf{A}$ *then* $\models^{\mathcal{K}} \Box \mathbf{A}$.

Corollary 2. *If* $\models \mathbf{A}$ *then* $\models \Box \mathbf{A}$.

Theorem 2.2. $\models \Box(\mathbf{A} \to \mathbf{B}) \to (\Box \mathbf{A} \to \Box \mathbf{B})$.

Proof. Select any $\mathcal{K} = \langle U, R \rangle$, $\mathcal{U}$ on $\mathcal{K}$, $u \in U$, and suppose $\models_u^{\mathcal{U}} \Box(\mathbf{A} \to \mathbf{B})$, $\models_u^{\mathcal{U}} \Box \mathbf{A}$. It will suffice by the truth definition to show that $\models_u^{\mathcal{U}} \Box \mathbf{B}$. So suppose uRt. Then $\models_t^{\mathcal{U}} \mathbf{A} \to \mathbf{B}$, $\models_t^{\mathcal{U}} \mathbf{A}$, by assumption. Hence $\models_t^{\mathcal{U}} \mathbf{B}$ by the truth definition. Thus $\models_u^{\mathcal{U}} \Box \mathbf{B}$, as was to be proved.

We may now define the system K. Consider the scheme

$A2$: $\quad \Box(\mathbf{A} \to \mathbf{B}) \to (\Box \mathbf{A} \to \Box \mathbf{B})$,

and the 'rule of necessitation'

RN: $\quad$ from $\mathbf{A}$ to derive $\Box \mathbf{A}$.

Specified axiomatically, K has as axioms any sentence of the form $A1$ or $A2$ and rules of inference MP and RN. Alternatively, if we think of *A2* as specifying the class of sentences of that form, we may define *K* as the smallest class of modal sentences Γ such that *A1* and *A2* are included in Γ and such that, if $\mathbf{A}, \mathbf{A} \to \mathbf{B} \in \Gamma$ then $\mathbf{B} \in \Gamma$, and if $\mathbf{A} \in \Gamma$ then $\square \mathbf{A} \in \Gamma$. Thus Γ is a system in the sense of Section 0.

Theorem 2.3. *If* $\vdash_K \mathbf{A}$ *then* $\vDash \mathbf{A}$.

Proof by induction on the length of **A**'s proof in *K*. It is clear from the truth definition that if $\mathbf{A} \in PC$ then $\vDash \mathbf{A}$. The result follows from the definition of *K* together with 2.1 (Corollary 2) and 2.2, if we note that if $\vDash \mathbf{A}$ and $\vDash \mathbf{A} \to \mathbf{B}$ then $\vDash \mathbf{B}$.

We shall say that a system *S* (more particularly, an extension of *K*) is *normal* iff whenever $\mathbf{A} \in S$ then $\square \mathbf{A} \in S$ (this conforms to the terminology of McKinsey–Tarski [1]).[19] Alternatively, *S* is normal iff *S* provides the rule of necessitation, in an obvious sense. Thus *K* is normal—though not all its extensions are, as we shall see.[20] However, most extensions of *K* which are familiar in the literature *are* normal; indeed, they are frequently, though not always, axiomatized in such a way that *RN* is taken as a rule of inference.

Theorem 2.4. *If S is a normal K-system, then*

(a) *if* $\vdash_S \mathbf{A} \to \mathbf{B}$ *then* $\vdash_S \square \mathbf{A} \to \square \mathbf{B}$;
(b) $\vdash_S \square(\mathbf{A} \wedge \mathbf{B}) \leftrightarrow \square \mathbf{A} \wedge \square \mathbf{B}$;
(c) $\vdash_S \square(\mathbf{A}_1 \wedge \cdots \wedge \mathbf{A}_n) \leftrightarrow \square \mathbf{A}_1 \wedge \cdots \wedge \square \mathbf{A}_n \ (n \geq 2)$;
(d) *if* $\vdash_S \mathbf{B}_1 \wedge \cdots \wedge \mathbf{B}_n \to \mathbf{A}$, *then* $\vdash_S \square \mathbf{B}_1 \wedge \cdots \wedge \square \mathbf{B}_n \to \square \mathbf{A} \ (n \geq 0)$.

Proof. (a) Suppose $\vdash_S \mathbf{A} \to \mathbf{B}$. Then $\vdash_S \square(\mathbf{A} \to \mathbf{B})$ since *S* is normal, whence $\vdash_S \square \mathbf{A} \to \square \mathbf{B}$ by *A2* (since *S* is a *K*-system) and *MP*.

(b) Since $\vdash_S \mathbf{A} \wedge \mathbf{B} \to \mathbf{A}$ and $\vdash_S \mathbf{A} \wedge \mathbf{B} \to \mathbf{B}$ by *A1*, we have $\vdash_S \square(\mathbf{A} \wedge \mathbf{B}) \to \square \mathbf{A}$, $\vdash_S \square(\mathbf{A} \wedge \mathbf{B}) \to \square \mathbf{B}$ by part (a). Hence $\vdash_S \square(\mathbf{A} \wedge \mathbf{B}) \to \square \mathbf{A} \wedge \square \mathbf{B}$ by *A1*. Conversely, $\vdash_S \mathbf{A} \to (\mathbf{B} \to \mathbf{A} \wedge \mathbf{B})$ by *A1* so that $\vdash_S \square \mathbf{A} \to \square(\mathbf{B} \to \mathbf{A} \wedge \mathbf{B})$ by part (a). This gives $\vdash_S \square \mathbf{A} \to$

---

[19] The reader should beware that Kripke's definition of normality, which has been adopted by some authors, differs from that of Lemmon: a logic is normal in the sense of Kripke if and only if it is normal in the sense of Lemmon and also contains *T*.

[20] This claim is not substantiated in this work; cf. n. 12. A non-normal extension of *S4* is exhibited in McKinsey–Tarski [1]. An analysis in Lemmon's spirit of that system is given in K. Segerberg, *An essay in classical modal logic*, Philosophical studies published by the Philosophical Society of Uppsala, vol. 13 (Uppsala, 1971).

($\Box \mathbf{B} \to \Box(\mathbf{A} \wedge \mathbf{B})$) by *A2* and *A1*, and so $\vdash_S \Box \mathbf{A} \wedge \Box \mathbf{B} \to \Box(\mathbf{A} \wedge \mathbf{B})$.

(c) By induction on n, using (b).

(d) For $n = 0$, the result holds since S is normal. For $n = 1$, use part (a). For $n \geq 2$, use (a), (c), and *A1*.

Theorem 2.5. *If S is a normal K-system, then*

(a) $\vdash_S \Box^n \mathbf{A}_1 \vee \cdots \vee \Box^n \mathbf{A}_k \to \Box^n (\mathbf{A}_1 \vee \cdots \vee \mathbf{A}_k)$;

(b) *if* $\vdash_S \mathbf{A} \wedge \Diamond^n \mathbf{B}_1 \wedge \cdots \wedge \Diamond^n \mathbf{B}_k \to \bot$ *then* $\vdash_S \Box \mathbf{A} \to \Box^{n+1}(-\mathbf{B}_1 \vee \cdots \vee -\mathbf{B}_k)$ $(k \geq 1, n \geq 0)$.

Proof. (a) By *A1*, $\vdash_S \mathbf{A}_i \to \mathbf{A}_1 \vee \cdots \vee \mathbf{A}_k$ $(1 \leq i \leq k)$, hence by 2.4(a) applied n times $\vdash_S \Box^n \mathbf{A}_i \to \Box(\mathbf{A}_1 \vee \cdots \vee \mathbf{A}_k)$ $(1 \leq i \leq k)$. The result follows by *A1*.

(b) Suppose $\vdash_S \mathbf{A} \wedge \Diamond^n \mathbf{B}_1 \wedge \cdots \wedge \Diamond^n \mathbf{B}_k \to \bot$. Then by *A1* and the definition of $\Diamond^n$, $\vdash_S \mathbf{A} \to \Box^n - \mathbf{B}_1 \vee \cdots \vee \Box^n - \mathbf{B}_k$. Now part (a) gives $\vdash_S \mathbf{A} \to \Box^n(-\mathbf{B}_1 \vee \cdots \vee -\mathbf{B}_k)$, and 2.4(a) gives the result.

The converse of 2.3 is what we are after for completeness, of course. We shall establish the existence of a particular world system $\mathcal{K}_K$ from whose special properties the result will be immediately forthcoming. However, we want to imitate the construction of $\mathcal{K}_K$ for a wide variety of systems, to obtain as much generality as possible. Indeed, for *any* consistent K-system S we define a modal structure $\mathcal{U}_S$ as follows. Let $\mathcal{U}_S$ be the set of all maximal consistent extensions of S; then, since S is assumed consistent, $\mathcal{U}_S$ is non-empty in virtue of Theorem 0.1. Further, for $u, t \in U_S$, we define

$$uR_S t \quad \text{iff} \quad \{\mathbf{A} : \Box \mathbf{A} \in u\} \subseteq t.$$

In other words, $uR_S t$ holds iff for all sentences $\mathbf{A}$ such that $\Box \mathbf{A}$ belongs to u $\mathbf{A}$ belongs to t. We also have

$$uR_S t \quad \text{iff} \quad \{\Diamond \mathbf{A} : \mathbf{A} \in t\} \subseteq u.$$

For suppose $uR_S t$, $\mathbf{A} \in t$, $\Diamond \mathbf{A} \notin u$. Then $-\Diamond \mathbf{A} \in u$ since u is complete, whence $\Box - \mathbf{A} \in u$ (compare *D6*), $-\mathbf{A} \in t$, a contradiction with the consistency of t. Conversely, suppose that, for all $\mathbf{A} \in t$, $\Diamond \mathbf{A} \in u$, and that $\Box \mathbf{A} \in u$, $\mathbf{A} \notin t$. Then $-\mathbf{A} \in t$ by the completeness of t, so that $\Diamond - \mathbf{A} \in u$, $-\Box \mathbf{A} \in u$, a contradiction with the consistency of u. We shall employ this second version of R_S frequently in what follows. Finally, we define φ thus:

$$\varphi_S(i) = \{t : \mathbf{P}_i \in t\},$$

so that $u \in \varphi_S(i)$ iff $\mathbf{P}_i \in u$. Now put $\mathcal{U}_S = \langle U_S, R_S, \varphi_S \rangle$. We may also put $\mathcal{K}_S = \langle U_S, R_S \rangle$, so that $\mathcal{U}_S$ is an m.s. on $\mathcal{K}_S$.

Now we prove two important results for what follows.

Theorem 2.6. *Let S be a consistent normal K-system. Then for any $u \in U_S$, $\Box A \in u$ iff, for all t such that uR_St, $A \in t$.*

Proof. Suppose $\Box A \in u$, for $u \in U_S$, and uR_St, i.e., $\{A : \Box A \in u\} \subseteq t$. Then $A \in t$. Conversely, suppose $A \in t$ for all t such that $\{A : \Box A \in u\} \subseteq t$. Then A belongs to all maximal consistent S-extensions of $\{A : \Box A \in u\}$. By 0.2 it follows that $\{A : \Box A \in u\} \vdash_S A$. This means that we have $B_1, \ldots, B_n$ $(n \geq 0)$ such that $\Box B_1, \ldots, \Box B_n \in u$ and

$$\vdash_S B_1 \wedge \cdots \wedge B_n \to A.$$

By 2.4(d) it follows that

$$\vdash_S \Box B_1 \wedge \cdots \wedge \Box B_n \to \Box A,$$

since S is normal. By properties of maximal consistent sets, it follows that $\Box A \in u$.

Theorem 2.7. *Let S be a consistent normal K-system. Then for $u, t \in U_S$*

$$uR_S^n t \quad \text{iff} \quad \{A : \Box^n A \in u\} \subseteq t.$$

Proof. By induction on n. Case $n = 0$: we need to show that $u = t$ iff $\{A : A \in u\} \subseteq t$, i.e., iff $u \subseteq t$, so that one half is trivial. Suppose then that $u \subseteq t$, $A \in t$, and yet $A \notin u$. Then $-A \in u$ by completeness of u, so that $-A \in t$, contradicting the consistency of t. This gives the basis for the induction.

Now suppose the result holds for n. Assume $uR_S^{n+1}t$, i.e., $(\exists t')$ $(uR_St' \wedge t'R_S^n t)$, and pick $\Box^{n+1} A \in u$. By definition of R_S, this gives $\Box^n A \in t'$. By the inductive hypothesis, given $t'R_S^n t$, it follows that $A \in t$. Conversely, suppose $\{A : \Box^{n+1} A \in u\} \subseteq t$, and consider the set

$$\Gamma = \{A : \Box A \in u\} \cup \{\Diamond^n A : A \in t\}.$$

We show first that Γ is S-consistent. For otherwise we have $\Gamma \vdash_S \bot$ which means that there are sentences $B_1, \ldots, B_l$ with $\Box B_1, \ldots, \Box B_l \in u$, and sentences $C_1, \ldots, C_k \in t$, such that

$$\vdash_S B_1 \wedge \cdots \wedge B_l \wedge \Diamond^n C_1 \wedge \cdots \wedge \Diamond^n C_k \to \bot.$$

By 2.5(b) and 2.4(c), this gives

$$\vdash_S \Box B_1 \wedge \cdots \wedge \Box B_l \to \Box^{n+1}(-C_1 \vee \cdots \vee -C_k).$$

Since u is an extension of S, it follows that $\Box^{n+1}(-\mathbf{C}_1 \vee \cdots \vee -\mathbf{C}_k) \in u$. We may now conclude that $-\mathbf{C}_1 \vee \cdots \vee -\mathbf{C}_k \in t$. But since t is maximal consistent, it follows that $-\mathbf{C}_i \in t$ for some i, $1 \leq i \leq k$, contradicting the consistency of t. Thus Γ is S-consistent. Hence by 0.1 we may find a maximal consistent S-extension of Γ, say t'. Since $\{\mathbf{A} : \Box\mathbf{A} \in u\} \subseteq \Gamma$, we have $uR_S t'$ by definition of R_S. Since $\{\Diamond^n \mathbf{A} : \mathbf{A} \in t\} \subseteq \Gamma$, $\{\Diamond^n \mathbf{A} : \mathbf{A} \in t\} \subseteq t'$, from which it follows that $\{\mathbf{A} : \Box^n \mathbf{A} \in t'\} \subseteq t$. Hence by the inductive hypothesis $t' R_S^n t$. These results yield $uR_S^{n+1} t$, and complete the proof.

Corollary. *For consistent normal K-systems S, $uR_S^n t$ iff $\{\Diamond^n \mathbf{A} : \mathbf{A} \in t\} \subseteq u$.*

We turn now to the basic property of $\mathcal{U}_S$ defined above.

Theorem 2.8. *Let S be a consistent normal K-system. Then for any $u \in U_S$,*

$$\vDash_u^{\mathcal{U}_S} \mathbf{A} \quad \text{iff} \quad \mathbf{A} \in u.$$

Proof by induction on $\mathbf{A}$. In case $\mathbf{A}$ is $\mathbf{P}_i$, $\vDash_u^{\mathcal{U}_S} \mathbf{A}$ iff $u \in \varphi_S(i)$ (by the truth definition) iff $\mathbf{P}_i \in u$ (by definition of φ_S) iff $\mathbf{A} \in u$. In case $\mathbf{A}$ is $\bot$, not $\vDash_u^{\mathcal{U}_S} \mathbf{A}$ by the truth definition; but also not $\mathbf{A} \in u$, since u is consistent. In case $\mathbf{A}$ is $\mathbf{B} \to \mathbf{C}$, $\vDash_u^{\mathcal{U}_S} \mathbf{A}$ iff if $\vDash_u^{\mathcal{U}_S} \mathbf{B}$ then $\vDash_u^{\mathcal{U}_S} \mathbf{C}$ (truth definition) iff if $\mathbf{B} \in u$ then $\mathbf{C} \in u$ (inductive hypothesis) iff $\mathbf{B} \to \mathbf{C} \in u$ (by properties of maximal consistent sets) iff $\mathbf{A} \in u$. In case $\mathbf{A}$ is $\Box\mathbf{B}$, $\vDash_u^{\mathcal{U}_S} \mathbf{A}$ iff $(\forall t)(uR_S t \to \vDash_t^{\mathcal{U}_S} \mathbf{B})$ (truth definition) iff $(\forall t)(uR_S t \to \mathbf{B} \in t)$ (inductive hypothesis) iff $\Box\mathbf{B} \in u$ (by Theorem 2.6) iff $\mathbf{A} \in u$. This completes the induction.

Corollary. *For any consistent normal K-system S, $\vdash_S \mathbf{A}$ iff for all $u \in U_S$ $\vDash_u^{\mathcal{U}_S} \mathbf{A}$, i.e., iff $\vDash^{\mathcal{U}_S} \mathbf{A}$.*

Proof. By the theorem, for all $u \in U_S$ $\vDash_u^{\mathcal{U}_S} \mathbf{A}$ iff for all $u \in U_S$ $\mathbf{A} \in u$, i.e., iff $\mathbf{A}$ belongs to all maximal consistent S-systems. But this holds iff $\vdash_S \mathbf{A}$, by the corollary to 0.2.

In the light of this corollary, we see that, for any consistent normal K-system S, $\mathcal{U}_S$ is an exact semantic counterpart to theoremhood of S: any theorem of S is true at all u in $\mathcal{U}_S$, and for any non-theorem there is a world u at which it is not true in $\mathcal{U}_S$. The completeness of K itself is now immediate. For K is certainly consistent by 2.3, since there are non-valid sentences (e.g., $\bot$), and by definition normal. Hence

Theorem 2.9. $\vdash_K \mathbf{A}$ *iff* $\vDash \mathbf{A}$.

Proof. One half is 2.3. Suppose then not $\vDash_K \mathbf{A}$. By the corollary to 2.8, we have $u \in U_K$ such that not $\vDash_u^{\mathcal{U}_S} \mathbf{A}$. Hence not $\vDash \mathbf{A}$.

The same proof incidentally demonstrates

Theorem 2.10. $\vdash_K \mathbf{A}$ iff $\vDash^{\mathcal{K}_K} \mathbf{A}$.

Thus theoremhood of K is equivalent to validity in $\mathcal{K}_K$: in this sense $\mathcal{K}_K$ is a *characteristic* world system for K.

We are actually, in this section, after bigger fish than a mere completeness result for K. We wish to consider also extensions of K which result from K by adding extra axioms or axiom-schemes. For example, we might consistently add to K as axiom $\Box \bot$ or even $\mathbf{P}_0$—bear in mind that K is formulated without any rule of variable-substitution, and throughout this discussion we shall assume that K-extensions are formulated with *MP* and *RN* as sole rules of inference. More commonly, however, we shall be adding *schemes* (such as $\Box \mathbf{A} \to \mathbf{A}$, $\Box \mathbf{A} \to \Box\Box \mathbf{A}$, $\Diamond\Box \mathbf{A} \to \mathbf{A}$, $\Diamond\Box \mathbf{A} \to \Box \mathbf{A}$, and the like), i.e., *classes* of sentences of the same form. It will be important for what follows to bear in mind that such a scheme may be thought of as *the class of substitution instances* of a single sentence; thus $\Box \mathbf{A} \to \mathbf{A}$, as a scheme, is identical with the class of substitution instances of $\Box \mathbf{P}_0 \to \mathbf{P}_0$ (or indeed $\Box \mathbf{P}_n \to \mathbf{P}_n$, for any n), i.e., the class of sentences obtainable from $\Box \mathbf{P}_0 \to \mathbf{P}_0$ by replacing $\mathbf{P}_0$ throughout by some sentence. For a given sentence $\mathbf{A}$, let $\Gamma(\mathbf{A})$ denote the class of its substitution instances. Then $\mathbf{A}' \in \Gamma(\mathbf{A})$ iff $\mathbf{A}'$ is a substitution instance of $\mathbf{A}$. By $K_{\Gamma(\mathbf{A})}$ we shall mean the system resulting from K by adding $\Gamma(\mathbf{A})$ as extra axioms. Thus if $\mathbf{A}$ is $\Box \mathbf{P}_0 \to \mathbf{P}_0$, $K_{\Gamma(\mathbf{A})}$ is, in fact, the system resulting from adding to K the *scheme* $\Box \mathbf{A} \to \mathbf{A}$.

We begin by associating with each modal sentence $\mathbf{A}$ a *characteristic condition* on u, as follows. (i) If $\mathbf{A}$ is a sentence letter $\mathbf{P}_i$, then the characteristic condition on u, $c_\mathbf{A}(u)$, is to be

$$u \in S_i.$$

(ii) If $\mathbf{A}$ is $\bot$, then $c_\mathbf{A}(u)$ is

$$u \neq u.$$

(iii) If $\mathbf{A}$ is $(\mathbf{B} \to \mathbf{C})$, then $c_\mathbf{A}(u)$ is

$$(c_\mathbf{B}(u) \to c_\mathbf{C}(u)).$$

(iv) If $\mathbf{A}$ is $\Box \mathbf{B}$, then $c_\mathbf{A}(u)$ is

$$(\forall t)(uRt \to c_\mathbf{B}(t))$$

(where t is, of course, a variable not already appearing in $c_\mathbf{B}(u)$, and $c_\mathbf{B}(t)$ is the result of replacing occurrences of u in $c_\mathbf{B}(u)$ by t).

For example, if **A** is $(\Box\Box\mathbf{P}_2 \to \Box\mathbf{P}_1)$, then $c_\mathbf{A}(u)$ is

$$((\forall t)(uRt \to (\forall t')(tRt' \to t' \in S_2)) \to (\forall t)(uRt \to t \in S_1)).$$

Then for any sentence **A**, containing let us say atomic sentences $\mathbf{P}_{k_1}, \ldots, \mathbf{P}_{k_n}$, $c_\mathbf{A}(u)$ is a formula of predicate calculus containing *set*-variables $S_{k_1}, \ldots, S_{k_n}$, and the *individual* variable u as its *sole* free variable. This formula has an obvious meaning in connection with a given $\mathcal{U} = \langle U, R, \varphi \rangle$: u will be a member of U, and S_n can be taken as $\varphi(n)$ (P_n). We accordingly say that $c_\mathbf{A}(u)$ *holds* for $\mathcal{U} = \langle U, R, \varphi \rangle$, $t \in U$, if $c_\mathbf{A}(u)$ comes out *true* when u is taken as t, the S_n as $\varphi(n)$. We also define: $\mathcal{U}$ *satisfies* $c_\mathbf{A}(u)$ iff $c_\mathbf{A}(u)$ holds for *all* $t \in U$.

It is worth noting that if **A** is $\Diamond\mathbf{B}$, then $c_\mathbf{A}(u)$ is equivalent to

$$(\exists t)(uRt \wedge c_\mathbf{B}(t)),$$

and, if **A** is $\Box^n\mathbf{B}$ or $\Diamond^n\mathbf{B}$, then $c_\mathbf{A}(u)$ is equivalent to

$$(\forall t)(uR^n t \to c_\mathbf{B}(t))$$

or

$$(\exists t)(uR^n t \wedge c_\mathbf{B}(t)),$$

respectively. We shall make use of these facts in writing $c_\mathbf{A}(u)$ for particular **A**.

For a given **A**, $c_\mathbf{A}(u)$ states exactly what is required for **A** to be true at u in $\mathcal{U}$ because its definition exactly parallels the truth definition. Thus the inductive proof of the next theorem should be obvious.

Theorem 2.11. *For any* $\mathcal{U} = \langle U, R, \varphi \rangle$, $t \in U$, $\vDash_t^{\mathcal{U}} \mathbf{A}$ *iff* $c_\mathbf{A}(u)$ *holds for* $\mathcal{U}, t$.

Corollary. *For any* $\mathcal{U}$, $\vDash^{\mathcal{U}} \mathbf{A}$ *iff* $\mathcal{U}$ *satisfies* $c_\mathbf{A}(u)$.

Then validity of **A** in $\mathcal{U}$ is equivalent to $\mathcal{U}$'s satisfaction of $c_\mathbf{A}(u)$.

Now let us consider the easy case where we extend K by adding a *single* axiom, say **A**: call the resulting system $K_\mathbf{A}$. We are going to show that theoremhood of $K_\mathbf{A}$ is equivalent to validity in all $\mathcal{U}$ satisfying $c_\mathbf{A}(u)$. One half is

Theorem 2.12. *If* $\vdash_{K_\mathbf{A}} \mathbf{B}$ *then* $\vDash^{\mathcal{U}} \mathbf{B}$ *for all* $\mathcal{U}$ *satisfying* $c_\mathbf{A}(u)$.

Proof by induction on the length of **B**'s proof in $K_\mathbf{A}$. The result holds trivially for axioms of K, and it holds for **A** itself by Theorem 2.11, Corollary. Both *MP* and *RN* preserve validity in any $\mathcal{U}$ (see 2.1), so the theorem is proved, since these are $K_\mathbf{A}$'s sole rules of inference.

Turning to completeness, we first establish

Theorem 2.13. *If S is a consistent normal K-system and $\vdash_S \mathbf{A}$, then $\mathcal{U}_S$ satisfies $c_\mathbf{A}(u)$.*

Proof. Suppose S is a consistent normal K-system, $\vdash_S \mathbf{A}$. Then $\models^{\mathcal{U}_S}\mathbf{A}$ by 2.8, Corollary, whence $\mathcal{U}_S$ satisfies $c_\mathbf{A}(u)$ by 2.11, Corollary.

Now we are ready for

Theorem 2.14. *If $\models^{\mathcal{U}} \mathbf{B}$ for all $\mathcal{U}$ satisfying $c_\mathbf{A}(u)$, then $\vdash_{K_\mathbf{A}} \mathbf{B}$.*

Proof. Suppose not $\vdash_{K_\mathbf{A}} \mathbf{B}$. Then $K_\mathbf{A}$ is consistent, obviously, and by definition normal since it provides RN. Since $\vdash_{K_\mathbf{A}} \mathbf{A}$, 2.13 applies, and $\mathcal{U}_S$ satisfies $c_\mathbf{A}(u)$. But by 2.8, Corollary, not $\models^{\mathcal{U}_S}\mathbf{B}$. This gives the result.

2.12 and 2.14 together give completeness results for all systems $K_\mathbf{A}$ with respect to $\mathcal{U}$ satisfying $c_\mathbf{A}(u)$. (Notice, by the way, that if $K_\mathbf{A}$ is inconsistent, so that $\vdash_{K_\mathbf{A}} \bot$, then there is *no* $\mathcal{U}$ satisfying $c_\mathbf{A}(u)$, i.e., $c_\mathbf{A}(u)$ is an *inconsistent* condition on $\mathcal{U}$.) By way of illustration, consider the sentences $\Box^n \bot$, and the resulting systems $K_{\Box^n \bot}$. The characteristic conditions for these sentences are

$$(\forall t)(uR^n t \to t \neq t)$$

or, more simply,

$$(\forall t) - uR^n t.$$

Our results give that $\vdash_{K_{\Box^n \bot}} \mathbf{A}$ iff $\models^{\mathcal{U}} \mathbf{A}$ for all $\mathcal{U} = \langle U, R, \varphi \rangle$ such that $(\forall u)(\forall t) - uR^n t$. Since this condition makes no mention of the $\varphi(i)$ belonging to $\mathcal{U}$, we have: $\vdash_{K_{\Box^n \bot}} \mathbf{A}$ iff $\models^{\mathcal{K}} \mathbf{A}$ for all $\mathcal{K} = \langle U, R \rangle$ such that $(\forall u)(\forall t) - uR^n t$. Since there *are* such $\mathcal{K}$, all these systems are consistent. Similarly, we learn that $\vdash_{K_{\mathbf{P}_n}} \mathbf{A}$ iff $\models^{\mathcal{U}} \mathbf{A}$ for all $\mathcal{U} = \langle U, R, \varphi \rangle$ such that $(\forall u)(u \in \varphi(n))$; again consistency follows.

With this behind us, we may turn to the rather harder case of adding schemes, or classes of substitution instances, to K. We need first a basic result concerning substitution instances. Let us think hereafter of a substitution instance $\mathbf{A}'$ of $\mathbf{A}$ as being determined by a complete set of substitutions $\mathbf{B}_i/\mathbf{P}_i$, where if $\mathbf{P}_i$ does not occur in $\mathbf{A}$ then $\mathbf{B}_i = \mathbf{P}_i$.

Theorem 2.15. *Let $\mathbf{A}'$ be a substitution instance of $\mathbf{A}$, with substitutions $\mathbf{B}_i/\mathbf{P}_i$, and for $\mathcal{U} = \langle U, R, \varphi \rangle$, put $\mathcal{U}' = \langle U, R, \varphi' \rangle$ where $\varphi'(i) = \{t : \models^{\mathcal{U}}_t \mathbf{B}_i\}$. Then for any $u \in U$*

$$\models^{\mathcal{U}'}_u \mathbf{A} \quad \textit{iff} \quad \models^{\mathcal{U}}_u \mathbf{A}'.$$

Proof. If **A** is $\mathbf{P}_i$, so that **A**′ is $\mathbf{B}_i$, we have $\vDash_u^{\mathcal{U}'} \mathbf{A}$ iff $u \in \varphi'(i)$ iff $\vDash_u^{\mathcal{U}} \mathbf{A}'$. The other cases of the induction are trivial.

Now let the modal sentences be *enumerated*, as $\mathbf{A}_1, \mathbf{A}_2, \ldots, \mathbf{A}_n, \ldots$. Then for given $\mathcal{U} = \langle U, R, \varphi \rangle$ we define a system $\Sigma_{\mathcal{U}}$ of subsets $T_1, T_2, \ldots, T_n, \ldots$ of U, by putting $T_n = \{t : \vDash_t^{\mathcal{U}} \mathbf{A}_n\}$. Thus $\Sigma_{\mathcal{U}}$ contains all subsets of U of the form $\{t : \vDash_t^{\mathcal{U}} \mathbf{A}\}$ for a given modal sentence **A**. ($\Sigma_{\mathcal{U}}$ will always contain U itself and the null class, and be closed under intersection, union, and complementation, so that it forms in an obvious way a Boolean algebra; but it need not contain *all* subsets of U.) We say that $\mathcal{U}' = \langle U, R, \varphi' \rangle$ is *substitution-related* to $\mathcal{U} = \langle U, R, \varphi \rangle$ iff, for each i, $\varphi'(i) \in \Sigma_{\mathcal{U}}$; for short, let us say that $\mathcal{U}'$ is *s-related* to $\mathcal{U}$ in this case. Thus $\mathcal{U}'$ is s-related to $\mathcal{U}$ if it has the same U and R and its sets P_j' are drawn from $\Sigma_{\mathcal{U}}$. The connection of this notion with that of a substitution instance should be clear. For a given $\mathcal{U}$ and **A**, any $\mathbf{A}' \in \Gamma(\mathbf{A})$ with substitutions $\mathbf{B}_i/\mathbf{P}_i$ determines $\mathcal{U}'$ s-related to $\mathcal{U}$ if we put $\varphi'(i) = \{t : \vDash_t^{\mathcal{U}} \mathbf{B}_i\}$—compare the statement of 2.16 in which this $\mathcal{U}'$ is s-related to $\mathcal{U}$; conversely, any $\mathcal{U}'$ s-related to $\mathcal{U}$ determines a substitution instance of **A** if we take the substitutions $\mathbf{B}_i/\mathbf{P}_i$, where $\mathbf{B}_i$ are the sentences such that $\varphi'(i) = \{t : \vDash_t^{\mathcal{U}} \mathbf{B}_i\}$ (these exist, since $\varphi'(i) \in \Sigma_{\mathcal{U}}$).

Let us say that $\mathcal{U}$ *strongly satisfies* $c_{\mathbf{A}}(u)$ iff any $\mathcal{U}'$ s-related to $\mathcal{U}$ satisfies $c_{\mathbf{A}}(u)$. Since $\mathcal{U}$ is trivially s-related to itself, if $\mathcal{U}$ strongly satisfies $c_{\mathbf{A}}(u)$ then it satisfies $c_{\mathbf{A}}(u)$. We have seen that theoremhood in $K_{\mathbf{A}}$ is equivalent to validity in all $\mathcal{U}$ satisfying $c_{\mathbf{A}}(u)$; we are going to prove now that theoremhood in $K_{\Gamma(\mathbf{A})}$ (K to which are added all substitution instances of **A**) is equivalent to validity in all $\mathcal{U}$ strongly satisfying $c_{\mathbf{A}}(u)$.

Theorem 2.16. *If $\mathcal{U}$ strongly satisfies $c_{\mathbf{A}}(u)$, then* $\vDash^{\mathcal{U}} \mathbf{A}'$ *for all* $\mathbf{A}' \in \Gamma(\mathbf{A})$.

Proof. Let $\mathcal{U} = \langle U, R, \varphi \rangle$ strongly satisfy $c_{\mathbf{A}}(u)$, and select $\mathbf{A}' \in \Gamma(\mathbf{A})$ with substitutions $\mathbf{B}_i/\mathbf{P}_i$. We define $\mathcal{U}'$ as in the statement of 2.16; namely, put $\varphi'(i) = \{t : \vDash_t^{\mathcal{U}} \mathbf{B}_i\}$. Then $\mathcal{U}'$ is s-related to $\mathcal{U}$, so that $\mathcal{U}'$ satisfies $c_{\mathbf{A}}(u)$. But this means that $\vDash^{\mathcal{U}'} \mathbf{A}$ by 2.11, Corollary. Hence $\vDash^{\mathcal{U}} \mathbf{A}'$ by 2.15.

Theorem 2.17. *If* $\vdash_{K_{\Gamma(\mathbf{A})}} \mathbf{B}$ *then* $\vDash^{\mathcal{U}} \mathbf{B}$ *for all $\mathcal{U}$ strongly satisfying* $c_{\mathbf{A}}(u)$.

Proof resembles that of 2.12, using 2.16 in place of 2.11.

Theorem 2.18. *If S is a consistent normal K-system and $\vdash_S \mathbf{A}'$ for all $\mathbf{A}' \in \Gamma(\mathbf{A})$, then $\mathcal{U}_S$ strongly satisfies $c_{\mathbf{A}}(u)$.*

Proof. Suppose S a consistent normal K-system such that $\vdash_S \mathbf{A}'$ for all $\mathbf{A}' \in \Gamma(\mathbf{A})$, and select $\mathcal{U}'$ s-related to $\mathcal{U}_S$. By the remarks preceding 2.16, $\mathcal{U}'$ determines a substitution instance $\mathbf{A}'$ of $\mathbf{A}$; namely, where $\varphi'(i) = \{t : \vDash_t^{\mathcal{U}_S} \mathbf{B}_i\}$, use $\mathbf{B}_i/\mathbf{P}_i$. Since this $\mathcal{U}'$ satisfies the hypothesis of 2.15, we conclude that for any $u \in U_S \vDash_u^{\mathcal{U}'} \mathbf{A}$ iff $\vDash_u^{\mathcal{U}_S} \mathbf{A}'$. 2.15, we conclude that for any $u = U_S \vDash_u^{\mathcal{U}'} \mathbf{A}$ iff $\vDash_u^{\mathcal{U}_S} \mathbf{A}'$. Since $\vdash_{K_{\Gamma(\mathbf{A})}} \mathbf{A}'$, we may apply Corollary 2.8, to conclude $\vDash^{\mathcal{U}_S} \mathbf{A}'$ and so $\vDash^{\mathcal{U}} \mathbf{A}$. By 2.11, Corollary, it follows that $\mathcal{U}'$ satisfies $c_\mathbf{A}(u)$. Thus $\mathcal{U}_S$ strongly satisfies $c_\mathbf{A}(u)$.

Theorem 2.19. *If $\vDash^{\mathcal{U}} \mathbf{B}$ for all $\mathcal{U}$·strongly satisfying $c_\mathbf{A}(u)$, then $\vdash_{K_{\Gamma(\mathbf{A})}} \mathbf{B}$.*

Proof is entirely analogous to that of 2.14, bearing in mind that $\vdash_{K_{\Gamma(\mathbf{A})}} \mathbf{A}'$ for all $\mathbf{A}' \in \Gamma(\mathbf{A})$ and using 2.18 in place of 2.13.

2.17 and 2.19 together give completeness results for all systems $K_{\Gamma(\mathbf{A})}$ with respect to $\mathcal{U}$ strongly satisfying $c_\mathbf{A}(u)$. Now it is also clear that adding any *finite* set of schemes, say $\Gamma(\mathbf{A}_1), \ldots, \Gamma(\mathbf{A}_n)$, presents no new problems. For this is equivalent simply to adding all substitution instances of a suitably chosen conjunction $\mathbf{A}'_1 \wedge \cdots \wedge \mathbf{A}'_n$ (suitably chosen, that is, by diversifying where necessary atomic sentences $\mathbf{P}_i$ occurring in the $\mathbf{A}_j$'s). It is worth noting, however, that even if we add *indefinitely* many schemes, say $\Gamma(\mathbf{A}_1), \ldots, \Gamma(\mathbf{A}_n), \ldots$, our completeness results may be extended in a quite natural way.

Thus put $\Phi = \Gamma(\mathbf{A}_1) \cup \cdots \cup \Gamma(\mathbf{A}_n) \cup \cdots$, i.e., Φ is the union of all $\Gamma(\mathbf{A}_i)$, and let K_Φ be the result of adding to K the set Φ of additional axioms. Then we can show

$\vdash_{K_\Phi} \mathbf{B}$ iff $\vDash^{\mathcal{U}} \mathbf{B}$ in all $\mathcal{U}$ such that for all i, $\mathcal{U}$ strongly satisfies $c_{\mathbf{A}_i}(u)$.

The proof of this from left to right is by induction on $\mathbf{B}$'s proof, and rests on the fact that, if $\mathbf{A}'_j \in \Gamma(\mathbf{A}_j)$ and for all i, $\mathcal{U}$ strongly satisfies $c_{\mathbf{A}_i}(u)$, then $\mathcal{U}$ strongly satisfies $c_{\mathbf{A}_j}(u)$ so that $\vDash^{\mathcal{U}} \mathbf{A}'_j$ by 2.16. Conversely, if not $\vdash_{K_\Phi} \mathbf{B}$, then K_Φ is consistent and by definition normal; thus not $\vDash^{\mathcal{U}_S} \mathbf{B}$ by 2.8, Corollary; that $\mathcal{U}_S$ strongly satisfies $c_{\mathbf{A}_i}(u)$ for all i now follows from 2.18, since $\vdash_{K_\Phi} \mathbf{A}'_j$ for all $\mathbf{A}'_j \in \Gamma(\mathbf{A}_j)$. From those remarks, it should be clear that our completeness results are 'additive' in the obvious sense, even in the infinite case. Similar remarks can be made concerning extensions of K formed by adding a mixture of axioms and axiom-schemes, if we bear in mind the content of 2.11, Corollary, and 2.13 as well as 2.16 and 2.18. We therefore have, implicit in our theorems, completeness results of a sort for a very wide variety of K-systems indeed. In many particular cases, however, we can find better results than these by showing the equivalence of theoremhood in a given S to validity in all *world systems* $\mathcal{H}$ satisfying some condition c, rather than to validity in all *modal structures* $\mathcal{U}$ satisfying, or strongly satisfying, some condition. But we postpone further discussion of these matters to Section 4.

Section 3

Before turning to further completeness results, we establish in this section the main properties of the system K introduced in the last. We demonstrate first some of the main theorem schemes and derived rules for the system. Then we show the system to be decidable by two rather different methods.

In the first place, Theory 2.4(a) already shows that K provides the rule RM:

RM: from $\mathbf{A} \to \mathbf{B}$ to derive $\Box \mathbf{A} \to \Box \mathbf{B}$.

It follows that K also provides the rule RE:

RE: from $\mathbf{A} \leftrightarrow \mathbf{B}$ to derive $\Box \mathbf{A} \leftrightarrow \Box \mathbf{B}$.

(See the discussion following Theorem 0.3). Hence K is classical in the sense of Section 0, so that by 0.3 it provides the full substitutivity of provable material equivalents. Indeed, in the light of 2.4(a), we may generalize to the conclusion that *any* normal K-system provides RM and RE, whence 0.3 applies.

Another rule which K provides, which we call RB after Becker,[21] is

RB: from $\mathbf{A} \to \mathbf{B}$ to derive $\Diamond \mathbf{A} \to \Diamond \mathbf{B}$.

For given $\mathbf{A} \to \mathbf{B}$, we deduce $-\mathbf{B} \to -\mathbf{A}$ by PC, $\Box - \mathbf{B} \to \Box - \mathbf{A}$ by RM, $-\Box - \mathbf{A} \to -\Box - \mathbf{B}$ by PC, which gives $\Diamond \mathbf{A} \to \Diamond \mathbf{B}$ by $D6$. This rule is very useful derivationally. It clearly also is provided by any normal K-system, by 2.4(a).

Now we may give a list of the principal theorem schemes of K. They remain theorem schemes, of course, for any extension of K. Where proofs are easy, they are omitted.

$T1$: $\Diamond(\mathbf{A} \vee \mathbf{B}) \leftrightarrow \Diamond \mathbf{A} \vee \Diamond \mathbf{B}$ (compare 2.4(b)).

$T2$: $\Diamond(\mathbf{A}_1 \vee \cdots \vee \mathbf{A}_n) \leftrightarrow \Diamond \mathbf{A}_1 \vee \cdots \vee \Diamond \mathbf{A}_n$ (compare 2.4(c)).

$T3$: $\Box \mathbf{A} \vee \Box \mathbf{B} \to \Box(\mathbf{A} \vee \mathbf{B})$ (use RM and PC).

$T4$: $\Box \mathbf{A}_1 \vee \cdots \vee \Box \mathbf{A}_n \to \Box(\mathbf{A}_1 \vee \cdots \vee \mathbf{A}_n)$ (compare 2.5(a)).

$T5$: $\Diamond(\mathbf{A} \wedge \mathbf{B}) \to \Diamond \mathbf{A} \wedge \Diamond \mathbf{B}$ (use RB and PC).

---

[21] O. Becker, not to be confused with A. Becker who was mentioned on p. 2.

T6: $\Diamond(\mathbf{A}_1 \wedge \cdots \wedge \mathbf{A}_n) \to \Diamond \mathbf{A}_1 \wedge \cdots \wedge \Diamond \mathbf{A}_n$.

T7: $\Box^n \mathbf{A} \to (\mathbf{B} \Rightarrow^n \mathbf{A})$ (use *PC* and *RM n* times).

T8: $\Box^n -\mathbf{A} \to (\mathbf{A} \Rightarrow^n \mathbf{B})$.

(*T7* and *T8* may be called the generalized paradoxes of strict implication.)

T9: $\Box^n \mathbf{A} \leftrightarrow (-\mathbf{A} \Rightarrow^n \mathbf{A})$.

T10: $\Box^n \mathbf{A} \leftrightarrow (\top \Rightarrow^n \mathbf{A})$.

T11: $(\mathbf{A} \Rightarrow^n \mathbf{B}) \leftrightarrow -\Diamond^n (\mathbf{A} \wedge -\mathbf{B})$.

T12: $(\mathbf{A} \Rightarrow^n \mathbf{B}) \to (\Box^n \mathbf{A} \to \Box^n \mathbf{B})$.

Proof by induction on *n*. $n = 0$ holds by *PC*. For the inductive step, use *RM*, *A2*, and *PC*.

T13: $(\mathbf{A} \Rightarrow^{n+m} \mathbf{B}) \to (\Box^n \mathbf{A} \Rightarrow^m \Box^n \mathbf{B})$ (use *T12* and *RM m* times).

Rather more generally, we can easily show that, if $\mathbf{A} \to (\mathbf{B} \to \mathbf{C})$ is a tautology, then $\vdash_K \Box^{n+m} \mathbf{A} \to (\Box^n \mathbf{B} \Rightarrow^m \Box^n \mathbf{C})$.

T14: $\Box(\mathbf{A} \to \mathbf{B}) \to (\Diamond \mathbf{A} \to \Diamond \mathbf{B})$ (from *A2* by contraposition and *D6*).

T15: $(\mathbf{A} \Rightarrow^{n+m} \mathbf{B}) \to (\Diamond^n \mathbf{A} \Rightarrow^m \Diamond^n \mathbf{B})$.

T16: $\Box \mathbf{A} \wedge \Diamond \mathbf{B} \to \Diamond(\mathbf{A} \wedge \mathbf{B})$.

Proof. By $A2, \vdash_K -(\Box \mathbf{A} \to \Box -\mathbf{B}) \to -\Box(\mathbf{A} \to -\mathbf{B})$. The result follows by *PC*, *D6*, and the substitutivity of provable equivalents.

T17: $\Box^n \mathbf{A} \wedge \Diamond^n \mathbf{B} \to \Diamond^n (\mathbf{A} \wedge \mathbf{B})$ (use *T12* contraposed).

T18: $\Box(\mathbf{A} \vee \mathbf{B}) \to \Box \mathbf{A} \vee \Diamond \mathbf{B}$ (use *T16* contraposed and *PC*).

T19: $\Box^n (\mathbf{A} \vee \mathbf{B}) \to \Box^n \mathbf{A} \vee \Diamond^n \mathbf{B}$.

T20: $(\mathbf{A} \Leftrightarrow^{n+m} \mathbf{B}) \to (\Box^n \mathbf{A} \Leftrightarrow^m \Box^n \mathbf{B})$.

T21: $(\mathbf{A} \Leftrightarrow^{n+m} \mathbf{B}) \to (\Diamond^n \mathbf{A} \Leftrightarrow^m \Diamond^n \mathbf{B})$.

T22: $\Diamond^m \top \to \Diamond^n \top$, for $m \geq n$.

Proof. By *PC*, $\Diamond \top \to \top$, whence $\Diamond^2 \top \to \Diamond \top$, $\Diamond^3 \top \to \Diamond^2 \top$, etc., by *RB*. The result follows by *PC*.

T23: $\Box^n \bot \to \Box^m \bot$, for $m \geq n$.

T24: $\top \leftrightarrow \Box^n \top$ (since $\vdash_K \Box^n \top$ by *RN n* times).

T25: $\bot \leftrightarrow \Diamond^n \bot$.

T1–T25 should give the reader some idea of what turns up as theorems of K. Of course, any result here stated for $\to$ as main connective may be strengthened to a result with $\Rightarrow^n$ by RN n times. It is convenient to state them in the above form, however, since they mostly turn out to be theorems of far weaker systems than K (such as K^*, to be discussed later);[22] indeed, the only two schemes above which depend essentially on RN are T24 and T25; the rest may be obtained from A1 and A2 using RM instead. The analogy (to be expected, in view of the truth definition) between $\square$ and the universal quantifier, $\Diamond$ and the existential quantifier, should be clear, but it is not total. The reader will gain a better understanding of what is *not* in K when we turn to systems stronger than it (in the following two sections).

We now set about showing that K is decidable. As in the matter of completeness, our method is applicable also to many K-systems other than K itself, so we shall state our results in as general a form as possible, ready for future use. In fact we shall demonstrate that, for any non-theorem **A** of K, there exists a *finite* world system $\mathcal{K}$, an upper bound on whose size is determined by **A** itself, in which $-\mathbf{A}$ is satisfiable. It is clearly a mechanical test to determine whether $-\mathbf{A}$ is or is not satisfiable in all world systems up to a certain size, so that K's decidability follows. The upper bound on the size of $\mathcal{K}$ is determined by the *structure* of **A**.

We say that **B** is a *subformula* of **A** iff **B** is a modal sentence occurring in **A** (**A** is a subformula of itself). Let $\Gamma_\mathbf{A}$ be the set of subformulas of **A** ($\Gamma_\mathbf{A}$ will always be a finite set). For any consistent K-system S, and sentence **A**, we define an equivalence relation $\equiv$ on U_S, the set of all maximal consistent S-extensions, as follows:

$$\text{for } u, t \in U_S, u \equiv t \quad \text{iff, for all} \quad \mathbf{B} \in \Gamma_\mathbf{A}, \quad \mathbf{B} \in u \quad \text{iff} \quad \mathbf{B} \in t.$$

(That $\equiv$ *is* an equivalence relation should be obvious.) Now $\equiv$ partitions U_S into at most 2^n equivalence classes, where n is the cardinality of $\Gamma_\mathbf{A}$, i.e., the number of subformulas of **A**. For there are at most 2^n ways in which any $u, t \in U_S$ can be distinguished by $\equiv$ into different equivalence classes, corresponding to the 2^n subsets of $\Gamma_\mathbf{A}$. We put $[u] = \{t : u \equiv t\}$, and $\bar{U}_S$ the set of equivalence classes $[u]$ for $u \in U_S$. We also define a relation $\bar{R}_S$ between equivalence

---

[22] There is some uncertainty as to whether in the manuscript the superscript of K^* really is an asterisk. In any case n. 12 applies: all modal systems discussed in this work are extensions of K. (K^*—if that is the correct reading—might be identical with Lemmon's own system $C2$; see Lemmon [3]).

classes, as follows:

for $[u], [t] \in \bar{U}_S$, $[u]\bar{R}_S(t)$ iff, for all **B**,

if $\Box \mathbf{B} \in \Gamma_\mathbf{A}$ and $\Box \mathbf{B} \in u$ then $\mathbf{B} \in t$.

This should be compared with the definition of R_S in the last section:

for $u, t \in U_S$, $uR_S t$ iff, for all **B**,

if $\Box \mathbf{B} \in u$ then $\mathbf{B} \in t$.

Let us first see that this definition is correct—that the choice of representatives u, t from the equivalence classes $[u]$, $[t]$ is irrelevant to the truth or falsity of $[u]\bar{R}_S[t]$. Suppose, then, that $u \equiv u'$, $t \equiv t'$, and that for all **B** if $\Box \mathbf{B} \in \Gamma_\mathbf{A}$ and $\Box \mathbf{B} \in u$ then $\mathbf{B} \in t$. We need to show that for all **B** if $\Box \mathbf{B} \in \Gamma_\mathbf{A}$ and $\Box \mathbf{B} \in u'$ then $\mathbf{B} \in t'$, so suppose further that $\Box \mathbf{B} \in \Gamma_\mathbf{A}$, $\Box \mathbf{B} \in u'$. Then by definition of $\equiv$ it follows that $\Box \mathbf{B} \in u$. Hence $\mathbf{B} \in t$. But given $\Box \mathbf{B} \in \Gamma_\mathbf{A}$, i.e., $\Box \mathbf{B}$ is a subformula of **A**, it follows that $\mathbf{B} \in \Gamma_\mathbf{A}$ also. Thus $\mathbf{B} \in t'$ by definition of $\equiv$ again, as required.

Put $\bar{\mathcal{K}}_S = \langle \bar{U}_S, \bar{R}_S \rangle$ and $\bar{\mathcal{U}}_S = \langle \bar{U}_S, \bar{R}_S, \bar{\varphi}_S \rangle$, where

$$\bar{\varphi}_S(i) = \{[t] : \mathbf{P}_i \in \Gamma_\mathbf{A} \wedge \mathbf{P}_i \in t\}$$

(compare the definition of φ_S in Section 2). Note that if $uR_S t$ then $[u]\bar{R}_S[t]$, for $u, t \in U_S$. For suppose $uR_S t$, i.e., $\{\mathbf{A} : \Box \mathbf{A} \in u\} \subseteq t$, and $\Box \mathbf{B} \in \Gamma_\mathbf{A}$, $\Box \mathbf{B} \in u$. Then at once $\mathbf{B} \in t$.

Theorem 3.1. *Let S be a consistent normal K-system, and **A** any sentence; then for any $\mathbf{B} \in \Gamma_\mathbf{A}$, any $u \in U_S$,*

$$\vDash^{\bar{\mathcal{U}}_S}_{[u]} \mathbf{B} \quad iff \quad \vDash^{\mathcal{U}_S}_u \mathbf{B}.$$

Proof is by induction on the length of **B**. Consider the case $\mathbf{B} = \mathbf{P}_i$. Then $\vDash^{\bar{\mathcal{U}}_S}_{[u]} \mathbf{B}$ iff $[u] \in \bar{\varphi}_S(i)$ iff $\mathbf{P}_i \in \Gamma_\mathbf{A}$ and $\mathbf{P}_i \in u$, by definition of $\bar{\varphi}_S(i)$. But $\mathbf{P}_i \in \Gamma_\mathbf{A}$ anyway, so that this last is equivalent to $\mathbf{P}_i \in u$, which holds iff $\vDash^{\mathcal{U}_S}_u \mathbf{P}_i$ by definition of the φ_S of $\mathcal{U}_S$. The case where $\mathbf{B} = \bot$ is trivial, as is the case where **B** has the form $\mathbf{C} \to \mathbf{D}$. So suppose finally that $\mathbf{B} = \Box \mathbf{C}$. Further suppose $\vDash^{\bar{\mathcal{U}}_S}_{[u]} \mathbf{B}$, i.e., for all $[t]$ such that $[u]\bar{R}_S[t] \vDash^{\bar{\mathcal{U}}_S}_{[t]} \mathbf{C}$. To show $\vDash^{\mathcal{U}_S}_u \Box \mathbf{C}$, assume $uR_S t$. Then $[u]\bar{R}_S[t]$ by the remarks preceding this theorem, so that $\vDash^{\bar{\mathcal{U}}_S}_{[t]} \mathbf{C}$. The inductive hypothesis applies since if $\mathbf{B} \in \Gamma_\mathbf{A}$ then $\mathbf{C} \in \Gamma_\mathbf{A}$, and we conclude $\vDash^{\mathcal{U}_S}_t \mathbf{C}$. This gives $\vDash^{\mathcal{U}_S}_u \mathbf{B}$. Conversely, suppose $\vDash^{\mathcal{U}_S}_u \mathbf{B}$; in order to show $\vDash^{\bar{\mathcal{U}}_S}_{[u]} \mathbf{B}$, assume $[u]\bar{R}_S[t]$. By 2.8, we know $\mathbf{B} \in u$, i.e., $\Box \mathbf{C} \in u$.

Given $\mathbf{B} = \Box \mathbf{C} \in \Gamma_{\mathbf{A}}$, we conclude $\mathbf{C} \in t$ by definition of $\bar{R}_S$. Using 2.6 again, we conclude $\vDash_t^{\mathcal{U}_S} \mathbf{C}$, whence $\vDash_{[t]}^{\bar{\mathcal{U}}_S} \mathbf{C}$ by the inductive hypothesis. This gives $\vDash_{[u]}^{\bar{\mathcal{U}}_S} \mathbf{B}$, and the induction is complete.

Inspection of the above proof shows that it may be generalized a little. Let us say that a relation $\bar{\bar{R}}$ on $\bar{U}_S$ is *suitable* iff whenever $uR_S t$ then $[u]\bar{\bar{R}}[t]$ and $\bar{\bar{R}} \subseteq \bar{R}_S$. Put $\bar{\bar{\mathcal{K}}}_S = \langle \bar{U}_S, \bar{\bar{R}} \rangle$, $\bar{\bar{\mathcal{U}}}_S = \langle \bar{U}_S, \bar{\bar{R}}, \bar{\varphi}_S \rangle$, and say that $\bar{\bar{\mathcal{K}}}_S$, $\bar{\bar{\mathcal{U}}}_S$ are *suitable* in case their $\bar{\bar{R}}$ is suitable. Then

Theorem 3.2. *Let S be a consistent normal K-system and $\bar{\bar{\mathcal{U}}}_S$ suitable; then for any $\mathbf{B} \in \Gamma_{\mathbf{A}}$, any $u \in U_S$,*

$$\vDash_{[u]}^{\bar{\bar{\mathcal{U}}}_S} \mathbf{B} \quad \text{iff} \quad \vDash_u^{\mathcal{U}_S} \mathbf{B}.$$

Details of the proof are left to the reader. Decidability of K is at once forthcoming from 3.1.

Theorem 3.3. *If not $\vdash_K \mathbf{A}$ and $\mathbf{A}$ has n subformulas, then there exists a world system $\mathcal{K}$ with not more than 2^n worlds such that $-\mathbf{A}$ is satisfiable in $\mathcal{K}$.*

Proof. Let $\mathbf{A}$ have n subformulas, and suppose not $\vdash_K \mathbf{A}$. By 2.8, Corollary, we have $u \in U_S$ such that not $\vDash_u^{\mathcal{U}_S} \mathbf{A}$. $\Gamma_{\mathbf{A}}$ now determines $\bar{\mathcal{U}}_S$ on $\bar{\mathcal{K}}_S$, with not more than 2^n worlds, such that not $\vDash_{[u]}^{\bar{\mathcal{U}}_S} \mathbf{A}$ by 3.1. Thus $-\mathbf{A}$ is satisfiable in $\bar{\mathcal{K}}_S$.

Corollary 1. *K is decidable.*

Corollary 2. $\vdash_K \mathbf{A}$ *iff* $\vDash^{\mathcal{K}} \mathbf{A}$ *for all finite $\mathcal{K}$.*

Although this gives us the decidability of K, the decision procedure involved is hardly a practical one, at least for complex $\mathbf{A}$, since the number of distinct $\mathcal{K}$ with at most 2^n worlds grows very rapidly as n grows. We give, therefore, an alternative decision procedure which is far more useful in practice. However, we proceed rather obliquely, since there are other properties of K which we wish to reveal on the way.

Let us say that a system S *provides the rule of disjunction* iff whenever $\vdash_S \Box \mathbf{A}_1 \vee \cdots \vee \Box \mathbf{A}_n$ then $\vdash_S \mathbf{A}_i$ for some i $(1 \leq i \leq n)$.

(Unlike rules such as *RM*, *RB*, discussed earlier, this rule can hardly be used as a rule of *derivation*, since we would not in general know *which* $\mathbf{A}_i$ to derive from the disjunction. The rule should rather be compared to the well-known fact about intuitionistic logic (*IL*) that if $\vdash_{IL} \mathbf{A} \vee \mathbf{B}$ then either $\vdash_{IL} \mathbf{A}$ or $\vdash_{IL} \mathbf{B}$, which does not hold good for classical logic, as the law of excluded middle shows.) We shall show that K provides the rule of disjunction, and we shall find later that some extensions of K, though by no means all, also provide it. We begin, however, with a general result, relating S's providing the rule of disjunction to a certain property of $\mathcal{K}_S$.

A w.s. $\mathcal{K} = \langle U, R \rangle$ is *left-directed* (*l-directed*) iff for all $t_1, \ldots, t_n \in U$ there exists $u \in U$ such that $uRt_1, \ldots, uRt_n$ (for any n). This notion has a very plain graphical interpretation: $\mathcal{K}$ is l-directed iff for any finite set T of worlds of U there exists a world which leads by R to each member of T. We show

Theorem 3.4. *Let S be a consistent normal K-system; then S provides the rule of disjunction iff $\mathcal{K}_S$ is l-directed.*

Proof. Select S a consistent normal K-system. Suppose first that S provides the rule of disjunction, and pick any $t_1, \ldots, t_n \in U_S$. Consider the set

$$\Gamma = \{\Diamond \mathbf{A} : \mathbf{A} \in t_1\} \cup \cdots \cup \{\Diamond \mathbf{A} : \mathbf{A} \in t_n\}.$$

We show that Γ is S-consistent. For otherwise, $\Gamma \vdash_S \bot$, i.e., there are sentences $\mathbf{A}_1, \ldots, \mathbf{A}_m$ such that for any i ($1 \le i \le m$) $\mathbf{A}_i \in t_j$ for some j ($1 \le j \le n$) and

$$\vdash_S \Diamond \mathbf{A}_1 \wedge \cdots \wedge \Diamond \mathbf{A}_m \to \bot,$$

whence

$$\vdash_S \Box - \mathbf{A}_1 \vee \cdots \vee \Box - \mathbf{A}_m,$$

by a simple transformation. Since S provides the rule of disjunction, it follows that $\vdash_S - \mathbf{A}_i$ for some i. But then $-\mathbf{A}_i \in t_j$ for any t_j, contradicting the consistency of one of the t_j to which $\mathbf{A}_i$ actually belongs. Select u, therefore, a maximal consistent S-extension of Γ by 0.1. Then $\{\Diamond \mathbf{A} : \mathbf{A} \in t_i\} \subseteq u$ for any i, i.e., $uR_S t_i$, and $\mathcal{K}_S$ is l-directed.

Conversely, suppose $\mathcal{K}_S$ l-directed, and select $\mathbf{A}_1, \ldots, \mathbf{A}_n$ such that not $\vdash_S \mathbf{A}_i$ for each i ($1 \le i \le n$). We shall show that not $\vdash_S \Box \mathbf{A}_1 \vee \cdots \vee \Box \mathbf{A}_n$, so that S provides the rule of disjunction. Since not $\vdash_S \mathbf{A}_i$ for each i, the sets $S \cup \{-\mathbf{A}_i\}$ are S-consistent for each i. Put t_i a maximal consistent S-extension of $S \cup \{-\mathbf{A}_i\}$ by 0.1, and select $u \in U_S$ such that $uR_S t_i$ by the l-directedness of $\mathcal{K}_S$. Now $\mathbf{A}_i \notin t_i$, from which we may conclude (since S is consistent and normal) that not $\vDash_{t_i}^{\mathcal{U}_S} \mathbf{A}_i$ by 2.8. Since $uR_S t_i$, not $\vDash_u^{\mathcal{U}_S} \Box \mathbf{A}_i$ for any i, and so not $\vDash_u^{\mathcal{U}_S} \Box \mathbf{A}_1 \vee \cdots \vee \Box \mathbf{A}_n$, in view of the truth definition. By 2.8, Corollary, it follows that not $\vdash_S \Box \mathbf{A}_1 \vee \cdots \vee \Box \mathbf{A}_n$. Thus S provides the rule of disjunction.

Now let us prove that K does indeed provide the rule of disjunction.

Theorem 3.5. *K provides the rule of disjunction.*

Proof. Suppose for $\mathbf{A}_1, \ldots, \mathbf{A}_n$ that not $\vdash_K \mathbf{A}_i$ for each i ($1 \leq i \leq n$). We shall show that not $\vdash_K \Box \mathbf{A}_1 \vee \cdots \vee \Box \mathbf{A}_n$ by constructing a suitable *safe extension* of $\mathcal{K}_K$ (compare Section 1, especially 1.2). Given not $\vdash_K \mathbf{A}_i$, the sets $K \cup \{-\mathbf{A}_i\}$ are consistent, and we have $t_i \in U_K$ maximal consistent K-extensions of these sets by 0.1. Now select $t^* \notin U_K$, and define $\mathcal{K}_K^*$, a safe extension of $\mathcal{K}_K$, as follows:

$$\mathcal{K}_K^* = \langle U_K \cup \{t^*\}, R_K \cup \{\langle t^*, t_i \rangle : 1 \leq i \leq n\} \rangle.$$

Thus $\mathcal{K}_K^*$ has one more world than $\mathcal{K}_K$, namely t^*, and its relation, say R_K^*, is R_K to which is added ordered pairs $\langle t^*, t_i \rangle$. Corresponding to $\mathcal{U}_K$ on $\mathcal{K}_K$, we have $\mathcal{U}_K^*$ with the same φ_K. Now $\mathbf{A}_i \notin t_i$, so that not $\vDash_{t_i}^{\mathcal{U}_K} \mathbf{A}_i$ by 2.8, whence not $\vDash_{t_i}^{\mathcal{U}_K^*} \mathbf{A}_i$ by 1.2. By the truth definition it follows that not $\vDash_{t^*}^{\mathcal{U}_K^*} \Box \mathbf{A}_1 \vee \cdots \vee \Box \mathbf{A}_n$, and so not $\vdash_K \Box \mathbf{A}_1 \vee \cdots \vee \Box \mathbf{A}_n$ by 2.3.

Corollary 1. $\vdash_K \Box \mathbf{A}_1 \vee \cdots \vee \Box \mathbf{A}_n$ *iff* $\vdash_K \mathbf{A}_i$ *for some* i ($1 \leq i \leq n$).

Proof. The converse of the theorem follows from the fact that K has RN.

Corollary 2. $\vdash_K \Box \mathbf{A} \vee \Box \mathbf{B}$ *iff either* $\vdash_K \mathbf{A}$ *or* $\vdash_K \mathbf{B}$.

Corollary 3. $\vdash_K \Box \mathbf{A}$ *iff* $\vdash_K \mathbf{A}$.

Proofs. These are merely the cases $n = 2, n = 1$ of Corollary 1, respectively.

Corollary 4. $\mathcal{K}_K$ *is l-directed.*

Proof by 3.4 and 3.3.

Actually, we can prove a result which is a little more general than 3.5, and which affords a rather swift decision procedure for K. It is, however, a bit complicated to state.

Theorem 3.6. *If* $\mathbf{B}$ *contains no modal operators (no occurrences of* $\Box$*) and* $\vdash_K \mathbf{B} \vee \Diamond \mathbf{C} \vee \Box \mathbf{C}_1 \vee \cdots \vee \Box \mathbf{C}_n$, *then either* $\vdash_K \mathbf{B}$ *or* $\vdash_K \mathbf{C} \vee \mathbf{C}_i$ *for some* i ($1 \leq i \leq n$).

Proof. Let $\mathbf{B}$ contain no modal operators and suppose neither $\vdash_K \mathbf{B}$ nor $\vdash_K \mathbf{C} \vee \mathbf{C}_i$ for any i ($1 \leq i \leq n$). By 0.1, we may find $t, t_1, \ldots, t_n \in U_K$ such that $-\mathbf{B} \in t$ and $-\mathbf{C} \in t_i$, $-\mathbf{C}_i \in t_i$ for each i, since the sets $K \cup \{-\mathbf{B}\}$, $K \cup \{-\mathbf{C}, -\mathbf{C}_i\}$ are consistent. $\mathcal{K}_K^*$ is defined as in the proof of 3.5, but now we define the φ of $\mathcal{U}_K^*$ to be φ_K^*, as follows:

$$\varphi_K^*(i) = \begin{cases} \varphi_K(i) \cup \{t^*\}, & \text{if } \mathbf{P}_i \in t, \\ \varphi_K(i), & \text{otherwise.} \end{cases}$$

In other words, $\mathcal{U}_K^*$ is now $\langle U_K \cup \{t^*\}, R_K \cup \{\langle t^*, t_i\rangle : 1 \leq i \leq n\}, \varphi_K^*\rangle$, for $t^* \notin U_K$. The reader is left to verify that the analogue of 1.2 still goes through: namely, for $u \in U_K$,

$$\vDash_u^{\mathcal{U}_K} \mathbf{A} \quad \text{iff} \quad \vDash_u^{\mathcal{U}_K^*} \mathbf{A}.$$

Since $-\mathbf{B} \in t$, $\vDash_t^{\mathcal{U}_K} -\mathbf{B}$; bearing in mind the construction of φ_K^* and the fact that $\mathbf{B}$ has no modal operators, we may conclude that $\vDash_{t^*}^{\mathcal{U}_K^*} -\mathbf{B}$. Further, since $-\mathbf{C} \in t_i$ for each i so that $\vDash_{t_i}^{\mathcal{U}_K} -\mathbf{C}$, then $\vDash_{t^*}^{\mathcal{U}_K^*} \square -\mathbf{C}$ by the construction of $\mathcal{K}_K^*$. Finally, as in the proof of 3.5, we find that not $\vDash_{t^*}^{\mathcal{U}_K^*} \square \mathbf{C}_1 \vee \cdots \vee \square \mathbf{C}_n$, since not $\vdash_K^{\mathcal{U}_K} \mathbf{C}_i$. Putting these results together, we discover that not $\vDash_{t^*}^{\mathcal{U}_K^*} \mathbf{B} \vee \Diamond \mathbf{C} \vee \square \mathbf{C}_1 \vee \cdots \vee \square \mathbf{C}_n$, and so not $\vdash_K \mathbf{B} \vee \mathbf{C} \vee \square \mathbf{C}_1 \vee \cdots \square \mathbf{C}_n$ by 2.3. This completes the proof.

Corollary 1. *If $\mathbf{B}$ contains no modal operators,* $\vdash_K \mathbf{B} \vee \Diamond \mathbf{C} \vee \square \mathbf{C}_1 \vee \cdots \vee \square \mathbf{C}_n$ *iff either* $\vdash_K \mathbf{B}$ *or* $\vdash_K \mathbf{C} \vee \mathbf{C}_i$ *for some i* $(1 \leq i \leq n)$.

Proof. The converse of the theorem follows by *PC*, *RN*, *T18*.

We shall also need special cases of this theorem in which either $\mathbf{B}$ or $\Diamond \mathbf{C}$ is omitted. Notice that $\vdash_K \bot \vee \mathbf{A}$ iff $\vdash_K \mathbf{A}$ and $\vdash_K \Diamond \bot \vee \mathbf{A}$ iff $\vdash_K \mathbf{A}$ (compare *T25*). Thus we may take $\mathbf{B}$ or $\mathbf{C}$ as $\bot$ and eliminate them; indeed, if we eliminate both we are back to *K*'s providing the rule of disjunction. This justifies

Corollary 2. $\vdash_K \Diamond \mathbf{C} \vee \square \mathbf{C}_1 \vee \cdots \vee \square \mathbf{C}_n$ *iff* $\vdash_K \mathbf{C} \vee \mathbf{C}_i$ *for some i* $(1 \leq i \leq n)$.

Corollary 3. *If $\mathbf{B}$ contains no modal operators,* $\vdash_K \mathbf{B} \vee \square \mathbf{C}_1 \vee \cdots \vee \square \mathbf{C}_n$ *iff either* $\vdash_K \mathbf{B}$ *or* $\vdash_K \mathbf{C}_i$ *for some i* $(1 \leq i \leq n)$.

A variant on the theorem is

Corollary 4. *If $\mathbf{B}$ contains no modal operators, then* $\vdash_K \mathbf{B} \wedge \square \mathbf{C} \to \square \mathbf{C}_1 \vee \cdots \vee \square \mathbf{C}_n$ *iff either* $\vdash_K -\mathbf{B}$ *or* $\vdash_K \mathbf{C} \to \mathbf{C}_i$ *for some i* $(1 \leq i \leq n)$.

Special cases of interest are

Corollary 5. $\vdash_K \Diamond \mathbf{A} \vee \square \mathbf{B}$ *iff* $\vdash_K \mathbf{A} \vee \mathbf{B}$.

(Corollary 2, $n = 1$.)

Corollary 6. $\vdash_K \square \mathbf{A} \to \square \mathbf{B}$ *iff* $\vdash_K \mathbf{A} \to \mathbf{B}$.

Corollary 7. $\vdash_K \square \mathbf{A} \leftrightarrow \square \mathbf{B}$ *iff* $\vdash_K \mathbf{A} \leftrightarrow \mathbf{B}$.

For the case in which there are no $\mathbf{C}_i$ we find a slight variation:

Corollary 8. *If $\mathbf{B}$ contains no modal operators,* $\vdash_K \mathbf{B} \vee \Diamond \mathbf{C}$ *iff* $\vdash_K \mathbf{B}$.

Proof modifies that of the theorem for the case $n = 0$; details are left to the reader. (Essentially, we introduce t^* but there are no additions to R_K in creating $\mathcal{K}_K^*$.)

Our new decision procedure for K is based on the fact that any modal sentence **A** can be transformed into an equivalent (in K) *conjunctive normal form*, each of whose conjuncts has the structure

$$\mathbf{B} \vee \Diamond \mathbf{C} \vee \Box \mathbf{C}_1 \vee \cdots \vee \Box \mathbf{C}_n,$$

in which **B** contains no modal operators (it being allowed that **B**, **C**, or all the $\mathbf{C}_i$ may be missing). Since the question of the theoremhood of **A** is equivalent to the question of the theoremhood of each of these conjuncts, by the various corollaries to 3.6, this question in turn is equivalent to the question of the theoremhood of sentences with no, or at least fewer, necessity symbols. Thus eventually, theoremhood or non-theoremhood can be settled by straightforward truth-table procedures.

For the basic reduction to normal forms, we proceed as follows. An occurrence of $\Box$ in **A** is *outermost in* **A** iff it does not lie within the scope of another occurrence of $\Box$ in **A** (here **A** is to be thought of in primitive notation). Thus in $-\Box-\Box\bot$, the first $\Box$ is outermost, the second not. An occurrence of a subformula **B** of **A** is *modally outermost in* **A** iff it begins with an outermost occurrence of $\Box$. A subformula **B** of **A** is a *modal part of* **A** iff **B** has a modally outermost occurrence in **A**. An occurrence of a sentence letter $\mathbf{P}_i$ in **A** is *modalized in* **A** iff it lies within the scope of an occurrence of $\Box$ in **A**; otherwise *unmodalized*. A sentence letter $\mathbf{P}_i$ is *unmodalized in* **A** iff $\mathbf{P}_i$ has an unmodalized occurrence in **A**. An *atom* of **A** is either an unmodalized sentence letter in **A** or a modal part of **A**.

We can now apply the usual techniques for finding conjunctive normal forms equivalent to **A** by taking as atoms either unmodalized sentence letters or modal parts of **A**. The result is a conjunction $\mathbf{A}'$ each of whose conjuncts has the form

$$\mathbf{A}_1 \vee \cdots \vee \mathbf{A}_m,$$

where $\mathbf{A}_i$ is an atom in the sense defined, either non-negated or negated. By regrouping and applying *T2*, we can find for such conjuncts an equivalent sentence (in K) of the form

$$\mathbf{B} \vee \Diamond \mathbf{C} \vee \Box \mathbf{C}_1 \vee \cdots \vee \Box \mathbf{C}_n,$$

where **B** contains no occurrences of $\Box$.

The procedure we have in mind here may be clearer from illustrations, which are of interest in their own right. Consider first the sentence $\mathbf{A} = \Box(\mathbf{P}_0 \to \mathbf{P}_1) \to \Box(\Box\mathbf{P}_0 \to \Box\mathbf{P}_1)$. Then $\vdash_K \mathbf{A}$ iff $\vdash_K (\mathbf{P}_0 \to \mathbf{P}_1) \to (\Box\mathbf{P}_0 \to \Box\mathbf{P}_1)$ by 3.6, Corollary 6. Hence $\vdash_K \mathbf{A}$ if either $\vdash_K -(\mathbf{P}_0 \to \mathbf{P}_1)$ or $\vdash_K \mathbf{P}_0 \to \mathbf{P}_1$ by 3.6, Corollary 4. Since neither $-(\mathbf{P}_0 \to \mathbf{P}_1)$ nor $\mathbf{P}_0 \to \mathbf{P}_1$ is tautologous, it is clear that neither are theorems of K. Hence $\mathbf{A}$ is not a K-theorem either. Next, consider the sentences $\mathbf{A}_n = \Box^n \mathbf{P}_0 \to \Box^{n+1} \mathbf{P}_0$. Using 3.6, Corollary 6, yet again, we have $\vdash_K \mathbf{A}_n$ iff $\vdash_K \mathbf{P}_0 \to \Box\mathbf{P}_0$, i.e., iff $\vdash_K -\mathbf{P}_0 \vee \Box\mathbf{P}_0$. By 3.6, Corollary 3, therefore, $\vdash_K \mathbf{A}_n$ iff either $\vdash_K -\mathbf{P}_0$ or $\vdash_K \mathbf{P}_0$. Since neither are K-theorems, not $\vdash_K \mathbf{A}_n$ for any n. We may readily generalize this argument to show that the sentences $\mathbf{A}_{m,n} = \Box^m \mathbf{P}_0 \to \Box^n \mathbf{P}_0$ are theorems of K iff $m = n$.

It will be seen that this decision procedure, by contrast to the former, gives rather rapid results for average sentences.

Section 4

The two best known, and most studied, systems of modal logic are the Lewis systems *S4* and *S5* mentioned in the Introduction. They are both extensions of *K*, as is also the system *T* mentioned there. Indeed *T* may be axiomatized very simply by adding to *K* the scheme, which we shall call *T*:

T: $\Box A \to A$.

S4 may be axiomatized by adding, in addition to *T*, the scheme which we shall call *4*:

4: $\Box A \to \Box\Box A$.

S5 may now be defined by adding to *S4* the scheme which we shall call *B* (after Brouwer, see Kripke [2]):

B: $\Diamond\Box A \to A$.

Alternatively, to obtain *S5* we may add directly to *T* the scheme to be called *E*:

E: $\Diamond\Box A \to \Box A$.

For it turns out that, as axiomatic additions to *T*, *E* is equivalent deductively to *4* and *B* taken together. (The result that these formulations do yield the Lewis systems is given in McKinsey–Tarski [1]; the case of *S4* was announced in Gödel [1].)

These systems are frequently taken as points of departure for defining others. For example, for the purposes of deontic logic (where $\Box$ is meant to be interpreted as 'it ought to be that' or 'it is obligatory that') logicians have felt *T* to be intuitively unsatisfactory, and have considered in its place the (weaker) scheme:

D: $\Box A \to \Diamond A$.

(See von Wright [1], Prior [2]). Various generalizations of *4* have also been used, for example:

4^n: $\Box^n A \to \Box^{n+1} A$

(where $4^1 = 4$ itself). Again, in connection with an extension of *S4* called *S4.2*, the scheme

G: $\Diamond\Box A \to \Box\Diamond A$

plays an axiomatic role (see Dummett–Lemmon [1]).

Completeness results for *T, S4, S5* were given by Kripke [2], using an approach by means of semantic tableaux, and the earlier result for *K* is implicit in Kripke's work. We want in this section to extend these results to as wide a variety of systems as possible. In this connection the scheme *G*, or rather a generalization of it, is of special importance, namely:

$$G': \quad \Diamond^m \Box^n \mathbf{A} \to \Box^p \Diamond^q \mathbf{A}.$$

For it is to be observed that *all* the special schemes enunciated so far in this section are particular cases of G'. Thus *T* is the case $m, p, q = 0, n = 1$; 4^n the case $m, q = 0, p = n+1$; *E* the case $q = 0, m, n, p = 1$; and so on. We shall begin by establishing a completeness result for G'. In a way that will become clear, this result alone gives a completeness result for the vast majority of *K*-systems discussed in the literature.

The naming of modal systems presents quite peculiar problems, since there is little uniformity among modal logicians on the matter. Here we adopt the policy, no doubt far from ideal, of naming axiom schemes (as *T, B, 4, 4^n*, etc.) in such a way that a system name can be formed by concatenating names of its axiom-schemes after the symbol *K*. Thus for schemes $Z_1, \ldots, Z_n$, $KZ_1 \cdots Z_n$ names the system which results from *K*(*A1*, *A2*, and sole rules *MP*, *RN*) by adding the schemes $Z_1, \ldots, Z_n$ as axiom schemes. In the light of this convention, such system names as *KB, K4, KE4, KTB4* should be readily understood: thus *KTB4* has extra axioms *T, B*, and *4*, and is in fact, as we mentioned earlier, a version of *S5*. Since *T* is such a common axiom scheme, it is also convenient to abbreviate *KT* simply to *T*. Then the system *T* itself is *KT*, i.e., *K* with extra axioms *T*; *T4* is *KT4*, i.e., the Lewis system *S4*—here, unfortunately, we depart from what is perhaps the most standard name in the literature, but we wish to reserve *S* as a *variable* ranging over systems. *TB* is *KTB*, *K* to which *T* and *B* are added, in fact the Brouwersche system of Kripke [2]. Since the scheme *D* is a common ingredient in all deontic systems, we similarly abbreviate *KD* to *D*; thus *D4* is *KD4*, *DB* is *KDB*, etc. Finally, since the combination *B4* is also of frequent occurrence, we abbreviate it to *5*. This odd convention has the outcome that *T5*, or *KTB4*, just is the Lewis system *S5*, *K5* is rather *KB4*, *D5* is *KDB4*—which turns out also to be *S5* = *T5*, as we shall see. Perhaps the most basic systems are *K, D, T, T4, TB, T5*. They form in fact the following inclusions:

$$K \subseteq D \subseteq T \subseteq T4 \subseteq T5, \qquad (1)$$

$$K \subseteq D \subseteq T \subseteq TB \subseteq T5. \qquad (2)$$

More graphically:

```
              T4
             /
K----D----T
             \      T5
              TB
```

Now consider the systems KG'. (G' only becomes a fixed axiom scheme for a given choice of m, n, p, q, of course.) We are going to show that theoremhood in KG' is equivalent to validity in all $\mathcal{K} = \langle U, R \rangle$ *of a special kind*, namely all $\mathcal{K}$ satisfying the following condition:

(g'): $(\forall u, t, t')(uR^m t \wedge uR^p t' \rightarrow (\exists t_0)(tR^n t_0 \wedge t'R^q t_0))$

(note that g', like G', contains the parameters m, n, p, q; it only becomes a definite condition when these are fixed).

We shall show first that any theorem of KG' is valid in any $\mathcal{K}$ satisfying g'. To this end, since the rules of derivation of KG' (RN and MP) preserve validity in any $\mathcal{K}$ (compare 2.1, Corollary 1) and since the axioms of K itself are valid in any $\mathcal{K}$ (2.3), it will clearly suffice to show

Theorem 4.1. $\vDash^{\mathcal{K}} G'$, for any $\mathcal{K}$ satisfying g'.

Proof. Select $\mathcal{K} = \langle U, R \rangle$ satisfying g'. We need to show $\vDash^{\mathcal{K}} \Diamond^m \Box^n \mathbf{A} \rightarrow \Box^p \Diamond^q \mathbf{A}$ for any $\mathbf{A}$, so select $\mathcal{U}$ on $\mathcal{K}, u \in U$, and suppose

$$\vDash_u^{\mathcal{U}} \Diamond^m \Box^n \mathbf{A}.$$

By the truth definition (see especially the derived clause (vii)), this gives some t such that $uR^m t$ and $\vDash_t^{\mathcal{U}} \Box^n \mathbf{A}$. In order to show $\vDash_u^{\mathcal{U}} \Box^p \Diamond^q \mathbf{A}$, assume $uR^p t'$ (compare (vi) of the truth definition). Since $\mathcal{K}$ satisfies g', we may now conclude that there exists t_0 such that $tR^n t_0, t'R^q t_0$. Since $tR^n t_0, \vDash_t^{\mathcal{U}} \Box^n \mathbf{A}$, it follows that $\vDash_{t_0}^{\mathcal{U}} \mathbf{A}$. Since $t'R^q t_0$, it follows that $\vDash_{t'}^{\mathcal{U}} \Diamond^q \mathbf{A}$. Hence

$$\vDash_u^{\mathcal{U}} \Box^p \Diamond^q \mathbf{A},$$

which completes the proof.

Corollary. *If* $\vdash_{KG'} \mathbf{A}$ *then* $\vDash^{\mathcal{K}} \mathbf{A}$ *for all $\mathcal{K}$ satisfying g'.*

The converse of this corollary gives completeness. But to show this it suffices to show that $\mathcal{K}_{KG'} = \langle U_{KG'}, R_{KG'} \rangle$, where $U_{KG'}$ is the set of maximal consistent extensions of KG', satisfies g'. For the systems KG' are certainly consistent by 4.1, since it is clear from the nature of g' that there exist $\mathcal{K}$ satisfying g' for any m, n, p, q. And by definition they are normal. Hence 2.8, Corollary, applies; if not $\vdash_{KG'} \mathbf{A}$, we may find $u \in U_{KG'}$ such that not $\vDash_u^{\mathcal{U}_{KG'}} \mathbf{A}$. If we show that $\mathcal{K}_{KG'}$ satisfies g', our proof of the converse of our last corollary will be achieved. In fact, we prove something rather stronger.

Theorem 4.2. *For any consistent normal K-system S, if S contains G' then $\mathcal{K}_S$ satisfies g'.*

Proof. Let S be any consistent normal K-system containing G' (i.e., such that $G' \subseteq S$). To show that $\mathcal{K}_S$ satisfies g', we need to show

$$(\forall u, t, t')(uR_S^m t \wedge uR_S^p t' \to (\exists t_0)(tR_S^n t_0 \wedge t'R_S^q t_0)).$$

For $u, t, t' \in U_S$, therefore, assume $uR_S^m t, uR_S^p t'$. By 2.7, $uR_S^i t$ iff $\{\mathbf{A} : \square^i \mathbf{A} \in u\} \subseteq t$, so we conclude $\{\mathbf{A} : \square^m \mathbf{A} \in u\} \subseteq t$, $\{\mathbf{A} : \square^p \mathbf{A} \in u\} \subseteq t'$. Now consider the set

$$\Gamma = \{\mathbf{A} : \square^n \mathbf{A} \in t\} \cup \{\mathbf{A} : \square^q \mathbf{A} \in t'\}.$$

We shall show that Γ is S-consistent. For otherwise there are sentences $\mathbf{B}_1, \ldots, \mathbf{B}_j, \mathbf{C}_1, \ldots, \mathbf{C}_k$ such that $\square^n \mathbf{B}_i \in t (1 \leq i \leq j)$, $\square^q \mathbf{C}_i \in t$ $(1 \leq i \leq k)$ such that $\vdash_S -(\mathbf{B} \wedge \mathbf{C})$, where $\mathbf{B} = \mathbf{B}_1 \wedge \cdots \wedge \mathbf{B}_j$, $\mathbf{C} = \mathbf{C}_1 \wedge \cdots \wedge \mathbf{C}_k$. Now $\vdash_K \square^n(\mathbf{A}_1 \wedge \cdots \wedge \mathbf{A}_m) \leftrightarrow \square^n \mathbf{A}_1 \wedge \cdots \wedge \square^n \mathbf{A}_m$ (compare 2.4(c)), so that $\square^n \mathbf{B} \in t$, $\square^q \mathbf{C} \in t'$ also. Given $\vdash_S -(\mathbf{B} \wedge \mathbf{C})$, $\vdash_S \mathbf{B} \to -\mathbf{C}$, whence $\vdash_S \square^n \mathbf{B} \to \square^n -\mathbf{C}$. Since t is an extension of S, $\square^n -\mathbf{C} \in t$. But $\{\mathbf{A} : \square^m \mathbf{A} \in u\} \subseteq t$, whence $\lozenge^m \square^n -\mathbf{C} \in u$. But $G' \subseteq S \subseteq u$, so that $\square^p \lozenge^q -\mathbf{C} \in u$. Given $\{\mathbf{A} : \square^p \mathbf{A} \in u\} \subseteq t'$, we conclude finally that $\lozenge^q -\mathbf{C} \in t'$, i.e., $-\square^q \mathbf{C} \in t'$, contradicting the consistency of t'. This shows that Γ is S-consistent, and so has a maximal consistent S-extension by 0.1, say t_0. By definition of Γ, $\{\mathbf{A} : \square^n \mathbf{A} \in t\} \subseteq t_0$, $\{\mathbf{A} : \square^q \mathbf{A} \in t'\} \subseteq t_0$; in other words, $tR_S^n t_0$, $t'R_S^q t_0$ by 2.7. Thus $\mathcal{K}_S$ satisfies g'.

Corollary. $\vdash_{KG'} \mathbf{A}$ *iff* $\vDash^{\mathcal{K}} \mathbf{A}$ *for all $\mathcal{K}$ satisfying g'.*

(Notice that KG' itself satisfies the conditions of 4.2.)

Let us now see what this result comes to for important special cases. In G' put $m, p, q = 0$, $n = 1$; this gives T, as we have seen. Bearing in mind that R^0 is by definition simply identity, we find that g' becomes

$$(\forall u, t, t')(u = t \wedge u = t' \to (\exists t_0)(tRt_0 \wedge t' = t_0)).$$

But this is equivalent to

$$(\forall u)uRu$$

by elementary reasoning. Thus g' here is simply the condition that R is reflexive! Let us say that $\mathcal{K} = \langle U, R \rangle$ is *reflexive* if its R is reflexive. Then the last corollary transforms in this case into

Theorem 4.3. $\vdash_T \mathbf{A}$ *iff* $\models \mathbf{A}$ *for all reflexive* $\mathcal{K}$.

(This result is in Kripke [2]).

Now consider 4, or more generally 4^n, $\Box^n \mathbf{A} \to \Box^{n+1} \mathbf{A}$. The corresponding version of g' is

$$(\forall u, t, t')(u = t \wedge uR^{n+1}t' \to (\exists t_0)(tR^n t_0 \wedge t' = t_0)).$$

But this is in turn equivalent to

$$(\forall u, t')(uR^{n+1}t' \to uR^n t').$$

Let us say that a relation with this property is *n-transitive*. If now $n = 1$, the condition transforms into

$$(\forall u, t, t')(uRt \wedge tRt' \to uRt'),$$

the condition for R's transitivity. Let us call $\mathcal{K}$ n-transitive (transitive) if its R is n-transitive (transitive). We have

Theorem 4.4. $\vdash_{K4^n} \mathbf{A}$ *iff* $\models^{\mathcal{K}} \mathbf{A}$ *for all n-transitive* $\mathcal{K}$.

Corollary. $\vdash_{K4} \mathbf{A}$ *iff* $\models \mathbf{A}$ *for all transitive* $\mathcal{K}$.

For the scheme B, where $m = 1$, $n = 1$, $p = 0$, $q = 0$, g' transforms into

$$(\forall u, t)(uRt \to tRu),$$

the condition for R's symmetry; the reader is left to verify this for himself. Calling $\mathcal{K}$ symmetric if its R is symmetric, we have accordingly

Theorem 4.5. $\vdash_{KB} \mathbf{A}$ *iff* $\models \mathbf{A}$ *for all symmetric* $\mathcal{K}$.

For the scheme E, where $m, n, p = 1$, $q = 0$, g' becomes equivalent to

$$(\forall u, t, t')(uRt \wedge uRt' \to tRt').$$

Let us call a relation *euclidean* if it satisfies this condition (it may be read: worlds related to the same world are related to each other):

there seems to be no standard term here. $\mathcal{K}$ is euclidean if its relation R is. This gives

Theorem 4.6. $\vdash_{KE} \mathbf{A}$ iff $\vDash^{\mathcal{K}} \mathbf{A}$ *for all euclidean* $\mathcal{K}$.

For the scheme D, where $m = 0$, $n = 1$, $p = 0$, $q = 1$, g' boils down simply to

$$(\forall u)(\exists t) u R t.$$

Let us say that R is *serial* if it satisfies this condition, and $\mathcal{K}$ serial if its R is. Then

Theorem 4.7. $\vdash_D \mathbf{A}$ iff $\vDash^{\mathcal{K}} \mathbf{A}$ *for all serial* $\mathcal{K}$.

Since G itself is just G' without the superscripts m, n, p, q, the right condition here—say (g)—is simply (g') without superscripts. Other cases of (g'), for particular m, n, p, q, we shall draw upon as we need them in the sequel.

These present completeness results, like those of Section 2, are 'additive', in the following sense. Consider two schemes G_1, G_2, particular cases of G' for fixed choices of m, n, p, q, and let g_1, g_2 be the corresponding versions of g'. The corollary to 4.2 gives completeness results for KG_1 and KG_2. But it is also clearly shown that

$$\vdash_{KG_1 G_2} \mathbf{A} \text{ iff } \vDash^{\mathcal{K}} \mathbf{A} \text{ for all } \mathcal{K} \text{ satisfying } g_1 \text{ and } g_2.$$

Details are left to the reader, but the essential step in the proof is to note that 4.2 is proved for any consistent normal K-systems S containing G' and not merely for particular systems KG'. All the completeness results we establish in fact will be additive in this manner, and we shall in future tacitly add them when we want to. This means that we can 'read off' from our theorems completeness results for the systems TB, $T4$, $T5$. Notice that if R in $\mathcal{K}$ is both reflexive and transitive, $\mathcal{K}$ is a *quasi-ordering*, and let us say that $\mathcal{K}$ is an *equivalence* if its R is an *equivalence relation* (reflexive, symmetric, and transitive). Then we have

Theorem 4.8. $\vdash_{TB} \mathbf{A}$ iff $\vDash^{\mathcal{K}} \mathbf{A}$ *for all reflexive symmetric* $\mathcal{K}$; $\vdash_{T4} \mathbf{A}$ iff $\vDash^{\mathcal{K}} \mathbf{A}$ *for all quasi-orderings* $\mathcal{K}$; $\vdash_{T5} \mathbf{A}$ iff $\vDash^{\mathcal{K}} \mathbf{A}$ *for all equivalences* $\mathcal{K}$.

(These results are again in Kripke [2].)

Our completeness results enable us to show quite a lot very simply (though non-constructively) about the comparative strength of various modal systems. We continue with some illustrations of this. In the first place,

(i) if R is reflexive, then R is serial

(uRu implies $(\exists t)uRt$). It follows that D is included in T, for any sentence valid in all serial $\mathcal{K}$ (theorem of D) is *a fortiori* valid in all reflexive $\mathcal{K}$ (theorem of T). Of course, this can also be shown constructively by proving D as a theorem scheme of T—a rather easy matter. This now justifies the earlier assertions (1) and (2), the other inclusions there being trivial.

More interestingly,

(ii) if R is reflexive, then R is euclidean iff R is both transitive and symmetric.

For suppose, for reflexive R, that R is euclidean, and suppose uRt; now uRu, whence tRu, given R euclidean, so that R is symmetric; now suppose uRt, tRt'; then tRu by R's symmetry, so that uRt' since R is euclidean; so R is transitive. Conversely, suppose R is an equivalence relation, and uRt, uRt'; then tRu by symmetry and tRt' by transitivity of R, so that R is euclidean. In fact, similar proofs show

(iii) if R is symmetric, R is transitive iff R is euclidean.

In virtue of (ii), the systems *TE* and *TB4* are identical, since their theorems are the sentences valid in exactly the same world systems; the system is just *T5* (Lewis' *S5*), and we have shown that in the presence of *T*, *E* is deductively equivalent to the union of *B* and *4* (a direct constructive proof is left to the reader), which is a claim we made earlier. (iii) additionally demonstrates that *B4* and *BE* are deductively equivalent combinations in *any* normal *K*-system. We introduced *5* as an abbreviation for *B4*, but we might as well have introduced it as an abbreviation for *BE*.

These considerations exhaust all possible ways of adding the schemes *B*, *4*, and *E* to the system *T*: only three distinct systems emerge, *TB*, *T4*, and *T5* = *TE*, for any pair of these is deductively equivalent here to any other pair and to all three. The situation is not so straightforward if we add directly to *K* or to *D*, however. Thus there turn out to be five distinct extensions of *K* of this form, which may be represented graphically as follows:

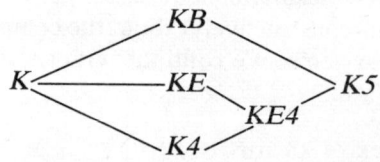

(thus $K \subseteq KB$, $K \subseteq KE$, $K \subseteq K4$, $KE \subseteq KE4$, $K4 \subseteq KE4$, $KB \subseteq K5$, $KE4 \subseteq K5$). That there are no more follows from the fact that $KB4 = KBE = K5$, as observed. When we add to D, the same pattern emerges with one slight exception: the system $D5$ turns out simply to be $T5$ again. This is a consequence of the fact that

(iv) if R is serial, transitive, and symmetric, then R is reflexive.

(We know that uRt for some t by R's serialness, whence tRu by symmetry, so that uRu by transitivity.) A direct proof that T is a theorem scheme of $D5$—not too easy—is left to the reader. Thus our pattern now is

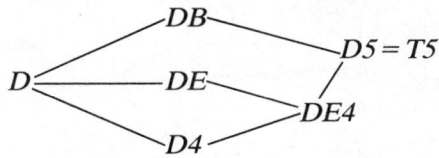

We have not yet substantiated the claim that these several systems are *distinct*, i.e., that the inclusions mentioned are *proper*. But this is usually not difficult, if we bear in mind once again the completeness results. For example, we can falsify $\Box \mathbf{P}_0 \to \mathbf{P}_0$ by a suitable $\mathcal{U}$ on any non-reflexive $\mathcal{K}$ at a world u such that $-uRu$; since $\mathcal{K}$ may well still be serial, we conclude that D is *properly* included in T. By way of further illustration, let us indicate how to show that $DE4$ is properly included in $D5 = T5$. Consider $\mathcal{K} = \langle \{u, t\}, \{\langle u, t\rangle, \langle t, t\rangle\}\rangle$. Diagrammatically, $\mathcal{K}$ may be represented in an obvious fashion as

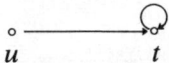

where the loop on t corresponds to the fact that tRt. It is easy to see that R is serial, euclidean, and transitive, though not symmetric. Thus any theorem of $DE4$ is valid in $\mathcal{K}$. But $\Diamond \Box \mathbf{P}_0 \to \mathbf{P}_0$ is falsified for $\mathcal{U}$ on $\mathcal{K}$ such that $\varphi(0) = \{t\}$, since for such $\mathcal{U} \vDash_u^\mathcal{U} \Diamond \Box \mathbf{P}_0$ but also $\vDash_u^\mathcal{U} -\mathbf{P}_0$. In a similar way, all our claimed inclusions can be shown to be proper (where they are not identities). In general, a claim of proper inclusion of two systems will thus follow from the exhibition of a world system $\mathcal{K}$ whose R has certain properties but not others. Entirely similar remarks may be made about demonstrations of the *independence* of systems S_1, S_2, i.e., the claim that neither $S_1 \subseteq S_2$ nor $S_2 \subseteq S_1$.

We find ourselves confronted, then, with fifteen distinct basic modal systems (including K) which result from K by adding D, T, B, E, and 4 in all possible ways. A graphical representation of them is the following:

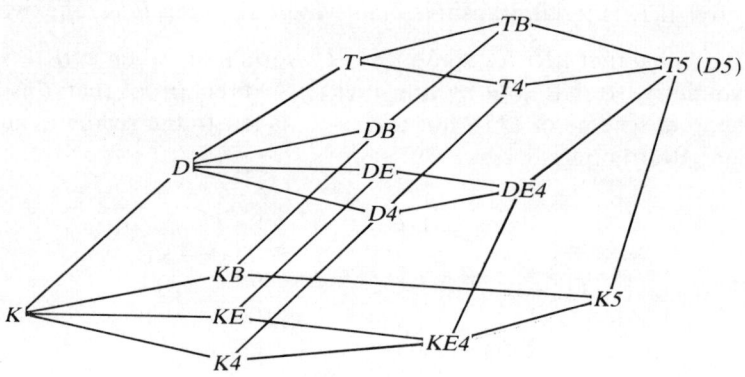

These may be considered the most fundamental systems of modal logic, although there are indefinitely many others intercalated with them for which we also have completeness results in the light of 4.2, Corollary.

For the record and for future reference, we now list the results of putting one or more of the parameters m, n, p, q in G' equal to 0, together with the resultant version of g' with respect to which the resultant scheme is complete. It is to be noted that by contraposition and obvious transformations G' is equivalent deductively to the scheme

$$\Diamond^p \Box^q \mathbf{A} \to \Box^m \Diamond^n \mathbf{A},$$

a fact which corresponds to the symmetry of g' with respect to the parameters m, p and n, q. Thus many of the subsequent cases collapse into each other.

A: the case $q = 0$ (equivalently, $n = 0$). The scheme is

$$\Diamond^m \Box^n \mathbf{A} \to \Box^p \mathbf{A}.$$

In this case g' transforms into

$$(\forall u, t, t')(uR^m t \land uR^p t' \to tR^n t').$$

Our completeness result for E is but a special case of this, and the condition given a generalized version of the euclidean condition.

Subcases. (i) Put $m = 0$. For the resulting scheme

$$\Box^n \mathbf{A} \to \Box^p \mathbf{A}$$

the corresponding completeness condition is

$$(\forall u, t')(uR^p t' \to uR^n t').$$

The schemes T, 4^n are special cases of this.
  (ii) Put $p = 0$. For the resulting scheme

$$\Diamond^m \Box^n \mathbf{A} \to \mathbf{A}$$

the corresponding condition is

$$(\forall u, t)(uR^m t \to tR^n u).$$

The scheme B is the special case: $m = n = 1$.
  (iii) Put $n = 0$. Then corresponding to the scheme

$$\Diamond^m \mathbf{A} \to \Box^p \mathbf{A}$$

we have the condition

$$(\forall u, t, t')(uR^m t \wedge uR^p t' \to t = t').$$

Notice that in the special case $m = p = 1$, the condition affirms that for any u there is *at most one* world to which it leads by R.

Each of these subcases in turn subdivide into two. Thus in (i) putting $p = 0$, we find:

𝔸(i)(a): Scheme: $\Box^n \mathbf{A} \to \mathbf{A}$; condition: $(\forall u)uR^n u$.

𝔸(i)(b): Scheme: $\mathbf{A} \to \Box^p \mathbf{A}$; condition: $(\forall u, t)(uR^p t \to u = t)$.

𝔸(ii)(a), (b) and 𝔸(iii)(a), (b) collapse back into these two cases.
  𝔹: the case $p = 0$ (equivalently, $m = 0$). The scheme is

$$\Diamond^m \Box^n \mathbf{A} \to \Diamond^q \mathbf{A}.$$

The corresponding condition is

$$(\forall u, t)(uR^m t \to (\exists t_0)(tR^n t_0 \wedge uR^q t_0)).$$

Subcases. (i) Put $m = 0$. For the scheme

$$\Box^n \mathbf{A} \to \Diamond^q \mathbf{A}$$

we find as corresponding condition

$$(\forall u)(\exists t_0)(uR^n t_0 \wedge uR^q t_0).$$

In case here $n = q$, we find corresponding to $\Box^n \mathbf{A} \to \Diamond^n \mathbf{A}$ the condition $(\forall u)(\exists t_0)uR^n t_0$. In case in addition $n > 0$, it is easy to see that this condition is equivalent to the condition $(\forall u)(\exists t)uRt$. Thus the schemes $\Box^n A \to \Diamond^n A$, for $n > 0$, are all deductively equivalent to D, the case $n = 1$; this can also be shown directly very easily. (In case $n = 0$, the condition collapses into a logical truth $(\forall u)(\exists t)(u = t)$, as we should expect.)

(ii) Put $n = 0$. We are back to A(i), contraposed.

(iii) Put $q = 0$. We are back to A(ii).

We conclude this section with two final remarks. First, consider the sentences $\Diamond^n T$. By the results of Section 2 (see 2.12, 2.14), $\vdash_{K \Diamond^n T} A$ iff $\vDash^{\mathcal{U}} A$ for all $\mathcal{U}$ satisfying $c_{\Diamond^n T}(u)$. Now this condition may be written equivalently

$$(\exists t)(u R^n t \wedge t = t).$$

From our remarks under B(i) above, it is clear that all the systems $K_{\Diamond^n T}$ are identical with each other and with the system D. Thus D may be axiomatized more simply by adding the simple sentence $\Diamond T$ as extra axiom to K in place of the scheme D; we shall utilize this fact in what follows. (A direct proof is not hard to find.) This point raises the whole question of the interrelationship between the rather trivial completeness results of Section 2 and our present ones, but we reserve discussion of this until the end of the next section.

Second, consider the scheme

$F: \quad \Diamond A \leftrightarrow \Box A$.

The resulting system KF is of some interest. Since F is deductively equivalent to the two schemes $\Diamond A \to \Box A$ and D (or the scheme $\Diamond A \to \Box A$ and the sentence $\Diamond T$), theoremhood in KF is equivalent to validity in all $\mathcal{K}$ whose R is serial and in addition satisfies the condition

$$(\forall u, t, t')(uRt \wedge uRt' \to t = t')$$

(compare A(iii) above). These conditions together are clearly equivalent to the condition

$$(\forall u)(\exists! t) uRt:$$

each world u leads to a *unique* world t. Let us say that R is *functional* in this case and that $\mathcal{K}$ with functional R is itself functional. Then

Theorem 4.9. $\vdash_{KF} A$ *iff* $\vDash^{\mathcal{K}} A$ *for all functional* $\mathcal{K}$.

(This is in disguise two special cases of 4.2, Corollary, combined.) A functional $\mathcal{K}$ may be equally well thought of as a system $\langle U, \delta \rangle$, where δ is a function from U into U (i.e., such that, for $u \in U$, $\delta(u) \in U$). For given functional $\mathcal{K} = \langle U, R \rangle$ we may define $\delta_R(u)$ as *the* t such that uRt, and conversely given a system $\langle U, \delta \rangle$ we may define an accessibility relation R_δ such that $uR_\delta t$ iff $\delta(u) = t$.

In terms of δ-*semantics*—semantics based on systems $\langle U, \delta \rangle$—our truth definition could have read

(iv′) $\vDash_u^{\mathcal{U}} \Box \mathbf{A}$ iff $\vDash_{\delta(u)}^{\mathcal{U}} \mathbf{A}$.

It would then have turned out, as we can easily see from 4.9, that the class of valid sentences in the new sense forms a system identical with *KF*. We shall have more to say about δ-semantics in Chapter III. We shall also discover an interpretational point to the system *KF* in connection with tense logic in Chapter II.[23]

---

[23] Chapters II and III were never written; cf. the editor's preface. An early attempt to explore the δ-semantics is found in G. H. von Wright, "'And next'", *Acta philosophica Fennica*, fasc. 18 (1965), pp. 275–303, and there is perhaps an anticipation of it in Prior [3]. See K. Segerberg, "On the logic of 'to-morrow'", *Theoria*, vol. 33 (1967), pp. 45–52, and Appendix B of A. N. Prior, *Past, present and future* (Oxford: Clarendon Press, 1967).

Section 5

Although, as we have seen, many of the most familiar systems of modal logic can be shown complete with respect to special classes of world systems $\mathcal{K}$ (in the sense that their theorems are exactly the sentences valid in those classes), there are others that cannot. We propose then in the first place to generalize considerably our result concerning G'. This will enable us to give completeness results of this kind for all normal K-systems which, to our knowledge, appear in the literature—with a notable exception, to be mentioned later.

We need some preliminary notions and results, which are of interest in their own right. By a *modality* is understood any finite sequence of the symbols $\Box$, $\Diamond$, and $-$ (including the null sequence, the null modality, which we denote by ϕ). Thus $\Box$, $\Diamond$, $\Box^2-\Diamond$, $-\Box^2-\Diamond^3-$ are all modalities. A modality is *affirmative* if it contains an even number of (or zero) occurrences of $-$; otherwise *negative*. Thus the null modality ϕ is affirmative, as are the first two and the last of the four just exhibited. For a given modality φ, the *dual* of φ is the modality which results from φ by interchanging $\Box$ and $\Diamond$ throughout φ. Thus the dual of the negative $\Box^2-\Diamond\Box\Diamond$ is $\Diamond^2-\Box\Diamond\Box$. Some simple facts about modalities are given in the next theorem.

Theorem 5.1. (*a*) *For any affirmative modality φ, there exists k such that*

$$\vdash_K \varphi \mathbf{A} \leftrightarrow \Box^{p_1}\Diamond^{q_1} \cdots \Box^{p_k}\Diamond^{q_k} \mathbf{A}.$$

(*b*) *For any negative modality φ, there exists an affirmative modality ψ such that $\vdash_K \varphi \mathbf{A} \leftrightarrow -\psi \mathbf{A}$.*

(*c*) *For any modality φ, let φ' be its dual; then $\vdash_K \varphi \mathbf{A} \leftrightarrow -\varphi'-\mathbf{A}$.*

(*d*) *For any affirmative modality φ*

$$\vdash_K \varphi \mathbf{A}_1 \vee \cdots \vee \varphi \mathbf{A}_k \to \varphi(\mathbf{A}_1 \vee \cdots \vee \mathbf{A}_k),$$

$$\vdash_K \varphi(\mathbf{A}_1 \wedge \cdots \wedge \mathbf{A}_k) \to \varphi \mathbf{A}_1 \wedge \cdots \wedge \varphi \mathbf{A}_k.$$

Proofs of (a), (b), (c) are easy. For (d), we use induction on the length of φ. In case φ is the null modality, the result holds by *PC*. For the inductive step, bear in mind that

$$\vdash_K \Box \mathbf{A}_1 \vee \cdots \vee \Box \mathbf{A}_k \to \Box(\mathbf{A}_1 \vee \cdots \vee \mathbf{A}_k),$$

$$\vdash_K \Diamond \mathbf{A}_1 \vee \cdots \vee \Diamond \mathbf{A}_k \to \Diamond(\mathbf{A}_1 \vee \cdots \vee \mathbf{A}_k),$$

and use *RM* and *RB*.

By 5.1(c), any affirmative modality φ is reduced to one in what may be thought of as *standard form*, determined by the numbers $p_1, q_1, \ldots, p_k, q_k$. Thus $\diamond^2 -\!- \square^2$ is equivalent to $\square^0 \diamond^2 \square^2 \diamond^0$, determined by the sequence $\langle 0, 2, 2, 0 \rangle$. Now consider the condition

$$(\forall t_1')(tR^{p_1}t_1' \to (\exists t_1)(t_1'R^{q_1}t_1 \wedge (\forall t_2')(t_1R^{p_2}t_2' \to \cdots$$
$$(\exists t_k)(t_k'R^{q_k}t_k \wedge uR^n t_k) \cdots))), \qquad (1)$$

which has for parameters u, t, n, as well as the numbers $p_1, q_1, \ldots, p_k, q_k$. In the light of our previous remark, we abbreviate (1) to

$$uR^{\varphi,n}t \qquad (2)$$

where φ is a suitable affirmative modality. Then $uR^{\varphi,n}t$ says that u, t, n are related to each other as in (1), in accordance with the parameters $p_1, \ldots, q_k$ yielded by φ. Part of the interest in this notion can be seen from the next theorem.

Theorem 5.2. *For any* $\mathcal{U} = \langle U, R, \varphi \rangle$, $u, t \in U$, *if* $\vDash_u^{\mathcal{U}} \square^n \mathbf{A}$ *and* $uR^{\varphi,n}t$ *then* $\vDash_t^{\mathcal{U}} \varphi \mathbf{A}$.

Proof. For given $\mathcal{U}, u, t$, suppose $\vDash_u^{\mathcal{U}} \square^n \mathbf{A}$, $uR^{\varphi,n}t$. Then $(\forall t)$ $(uR^n t \to \vDash_t^{\mathcal{U}} \mathbf{A})$. Hence from (1) by elementary reasoning:

$$(\forall t_1')(tR^{p_1}t_1' \to \cdots (\exists t_k)(t_k'R^{q_k}t_k \wedge \vDash_{t_k}^{\mathcal{U}} \mathbf{A}) \cdots).$$

By 5.1(a), giving the representation of φ, and the truth definition, we conclude that $\vDash_t^{\mathcal{U}} \varphi \mathbf{A}$.

If φ is the null modality, $p_1 = q_1 = \cdots = p_k = q_k = 0$, then (1) collapses into $uR^n t$, since $R^{p_1}, R^{q_1}, \ldots$ all become identity. Thus $uR^{\phi,n}t$ iff $uR^n t$. Accordingly, the next theorem can be seen as a generalization of 2.7, and similarly concerns R_S for a given consistent normal K-system S.

Theorem 5.3. *Let S be a consistent normal K-system. Then for $u, t \in U_S$, $uR_S^{\varphi,n}t$ iff $\{\varphi \mathbf{A} : \square^n \mathbf{A} \in u\} \subseteq t$, where φ is any affirmative modality.*

Proof by induction on the length of φ. Where $\varphi = \phi$, the result is just 2.7. Suppose the result holds for φ. It clearly suffices to show that it consequently holds for $\diamond \varphi$ and $\square \varphi$, since φ is assumed affirmative. We accordingly treat two cases:

(i) Assume $uR_S^{\diamond \varphi, n}t$, to show $\{\diamond \varphi \mathbf{A} : \square^n \mathbf{A} \in u\} \subseteq t$. Then by (1) and (2)

$$(\exists t_1)(tR_S t_1) \wedge uR_S^{\varphi,n}t_1).$$

Pick $\Box^n \mathbf{A} \in u$. Then given $uR_S^{\varphi,n}t_1$, by the inductive hypothesis, $\varphi \mathbf{A} \in t_1$, so that $\Diamond \varphi \mathbf{A} \in t$ by tR_St_1. Conversely, suppose $\{\Diamond \varphi \mathbf{A} : \Box^n \mathbf{A} \in u\} \subseteq t$. To show $uR_S^{\Diamond \varphi, n}t$, we show that S-consistency of

$$\Gamma = \{\mathbf{A} : \Box \mathbf{A} \in t\} \cup \{\varphi \mathbf{A} : \Box^n \mathbf{A} \in u\}.$$

For if Γ is S-inconsistent, we have sentences $\mathbf{B}_1, \ldots, \mathbf{B}_k, \mathbf{C}_1, \ldots, \mathbf{C}_l$ such that $\Box \mathbf{B}_i \in t, \Box^n \mathbf{C}_j \in u$ ($1 \leq i \leq k$, $1 \leq j \leq l$) and

$$\{\mathbf{B}_1, \ldots, \mathbf{B}_k, \varphi \mathbf{C}_1, \ldots, \varphi \mathbf{C}_l\} \vdash_S \bot.$$

Put $\mathbf{B} = \mathbf{B}_1 \wedge \cdots \wedge \mathbf{B}_k$, $\mathbf{C} = \mathbf{C}_1 \wedge \cdots \wedge \mathbf{C}_l$; then clearly $\Box \mathbf{B} \in t$, $\Box^n \mathbf{C} \in u$. Using 5.1(d) and bearing in mind that φ is affirmative, we conclude

$$\vdash_S \mathbf{B} \to -\varphi \mathbf{C}.$$

By RM and the fact that $\Box \mathbf{B} \in t$, we conclude that $\Box - \varphi \mathbf{C} \in t$. Since $\Box^n \mathbf{C} \in u$, by our original hypothesis $\Diamond \varphi \mathbf{C} \in t$, contradicting the consistency of t. Given Γ S-consistent, we find t_1 by 0.1 such that tR_St_1 and $\{\varphi \mathbf{A} : \Box^n \mathbf{A} \in u\} \subseteq t_1$. Hence $uR_S^{\varphi,n}t_1$ by the inductive hypothesis. This shows, by (1) and (2), that $uR_S^{\Diamond \varphi, n}t$.

(ii) Suppose first that $\{\Box \varphi \mathbf{A} : \Box^n \mathbf{A} \in u\} \subseteq t$. To show $uR_S^{\Box \varphi, n}t$, by (1) and (2) we must show

$$(\forall t_1)(tR_St_1 \to uR_S^{\varphi,n}t_1). \tag{3}$$

Suppose therefore tR_St_1; by the inductive hypothesis, we need now to demonstrate that $\{\varphi \mathbf{A} : \Box^n \mathbf{A} \in u\} \subseteq t_1$. Select $\Box^n \mathbf{A} \in u$. Then $\Box \varphi \mathbf{A} \in t$ by our original assumption, so that $\varphi \mathbf{A} \in t_1$ by tR_St_1. Conversely, suppose *not* $\{\Box \varphi \mathbf{A} : \Box^n \mathbf{A} \in u\} \subseteq t$. Then we can find $\mathbf{B}$ such that $\Box^n \mathbf{B} \in u$, $-\Box \varphi \mathbf{B} \in t$. Consider

$$\Gamma' = \{\mathbf{A} : \Box \mathbf{A} \in t\} \cup \{-\varphi \mathbf{B}\}.$$

If Γ' is S-inconsistent, we may find $\mathbf{C} = \mathbf{C}_1 \wedge \cdots \wedge \mathbf{C}_k$ such that $\Box \mathbf{C} \in t$ and

$$\vdash_S \mathbf{C} \to \varphi \mathbf{B}.$$

But then by RM $\Box \varphi \mathbf{B} \in t$, contradicting the consistency of t. This means that by 0.1 we have t_1, a maximal consistent extension of Γ', such that tR_St_1 and yet $\Box^n \mathbf{B} \in u$, $-\varphi \mathbf{B} \in t_1$. By the inductive hypothesis, not $uR_S^{\varphi,n}t_1$. Since this negates (3) above, we conclude that not $uR_S^{\Box \varphi, n}t$. Our proof is complete.

We are now ready to consider the scheme

$$H' : \Diamond^{m_1}\Box^{n_1}\mathbf{A}_1 \wedge \cdots \wedge \Diamond^{m_k}\Box^{n_k}\mathbf{A}_k$$
$$\to \Box^{p_1}\Diamond^{q_1}(\varphi_1^1 \mathbf{A}_1 \wedge \cdots \wedge \varphi_k^1 \mathbf{A}_k)$$
$$\vee \cdots \vee \Box^{p_l}\Diamond^{q_l}(\varphi_1^l \mathbf{A}_1 \wedge \cdots \wedge \varphi_k^l \mathbf{A}_k)$$

in which φ_i^j ($1 \leq i \leq k$, $1 \leq j \leq l$) are affirmative modalities. This is a scheme in the parameters $m_1, n_1, \ldots, m_k, n_k, p_1, q_1, \ldots, p_l, q_l, k, l$, as well as in the parameters implicit in each φ_i^j. For these fixed parameters, we shall show completeness of KH' with respect to the class of $\mathcal{K}$ whose R satisfies:

(h'): $(\forall u, t_1, \ldots, t_k)[uR^{m_1}t_1 \wedge \cdots \wedge uR^{m_k}t_k$

$\to (\forall t_1')(uR^{p_1}t_1' \to (\exists t_1'')(t_1'R^{q_1}t_1'' \wedge t_1 r^{\varphi_1^1, n_1}t_1''$

$\wedge \cdots \wedge t_k R^{\varphi_k^1, n_k}t_1'')) \vee \cdots \vee (\forall t_l')(uR^{p_l}t_l'$

$\to (\exists t_l'')(t_l'R^{q_l}t_l'' \wedge t_1 R^{\varphi_1^l, n_1}t_l''$

$\wedge \cdots \wedge t_k R^{\varphi_k^l, n_k}t_l''))]$

This rather nasty condition should be compared, symbol for symbol, with the scheme H', whose outline it closely follows. The reader may care to observe that, if all modalities φ_i^j are taken as ϕ and $k = 1$, $l = 1$, H' collapses into G' and h' into g'.

Our completeness proof as usual falls into two halves.

Theorem 5.4. $\models^{\mathcal{K}} H'$ *for any $\mathcal{K}$ satisfying h'.*

Proof. Select $\mathcal{K} = \langle U, R \rangle$ satisfying h', $\mathcal{U}$ on $\mathcal{K}$, $u \in U$, and suppose $\models_u^{\mathcal{U}} \diamondsuit^{m_i} \square^{n_i} \mathbf{A}_i$ for each i ($1 \leq i \leq k$). This yields $t_1, \ldots, t_k$ such that $\models_{t_i}^{\mathcal{U}} \square^{n_i} \mathbf{A}_i$ and $uR^{m_i}t_i$ for each i ($1 \leq i \leq k$), by the truth definition. By h', we now deduce the consequent of h', so that for some j

$(\forall t_j')(uR^{p_j}t_j' \to (\exists t_j'')(t_j'R^{q_j}t_j'' \wedge t_1 R^{\varphi_1^j, n_1}t_j'' \wedge \cdots \wedge t_k R^{\varphi_k^j, n_k}t_j''))$

($1 \leq j \leq l$). We aim to show $\models_u^{\mathcal{U}} \square^{p_j} \diamondsuit^{q_j}(\varphi_1^j \mathbf{A}_1 \wedge \cdots \wedge \varphi_k^j \mathbf{A}_k)$, giving the result. So suppose $uR^{p_j}t_j'$. This gives t_j'' such that $t_j'R^{q_j}t_j''$ and $t_i R^{\varphi_i^j, n_i} t_j''$ for each i ($1 \leq i \leq k$). Since $\models_{t_i}^{\mathcal{U}} \square^{n_i} \mathbf{A}_i$ for each i, it follows by 5.2 that $\models_{t_j''}^{\mathcal{U}} \varphi_i^j \mathbf{A}_i$ for each i. Given $t_j'R^{q_j}t_j''$, we find that $\models_{t_j'}^{\mathcal{U}} \diamondsuit^{q_j}(\varphi_1^j \mathbf{A}_1 \wedge \cdots \wedge \varphi_k^j \mathbf{A}_k)$. The theorem is thus proved.

Corollary. *If $\vdash_{KH'} \mathbf{A}$ then $\models^{\mathcal{K}} \mathbf{A}$ for all $\mathcal{K}$ satisfying h'.*

Theorem 5.5. *For any consistent normal K-system S, if S contains H', then $\mathcal{K}_S$ satisfies h'.*

Proof. Let S be a consistent normal K-system such that $H' \subseteq S$, and select $u, t_1, \ldots, t_k \in U_S$ such that $uR_S^{m_i}t_i$, i.e., $\{\mathbf{A} : \square^{m_i}\mathbf{A} \in u\} \subseteq t_i$, for each i ($1 \leq i \leq k$). In order to derive the consequent of h' for R_S, let us further assume $uR_S^{p_1}t'_1, \ldots, uR_S^{p_l}t'_l$, i.e., $\{\mathbf{A}: \square^{p_j}\mathbf{A} \in u\} \subseteq t'_j$ for each j ($1 \leq j \leq l$), and define

$$\Gamma_j = \{\mathbf{A}: \square^{q_j}\mathbf{A} \in t'_j\} \cup \{\varphi_1^j \mathbf{A}: \square^{n_1}\mathbf{A} \in t_1\}$$
$$\cup \cdots \cup \{\varphi_k^j \mathbf{A}: \square^{n_k}\mathbf{A} \in t_k\}$$

for each j ($1 \leq j \leq l$). We show first that for some j Γ_j is S-consistent. For suppose Γ_j is S-inconsistent for each j. Then for each j we may find sentences $\mathbf{B}^j$, $\mathbf{C}_1^j, \ldots, \mathbf{C}_k^j$ such that $\square^{q_j}\mathbf{B}^j \in t'_j$, $\square^{n_1}\mathbf{C}_1^j \in t_1, \ldots, \square^{n_k}\mathbf{C}_k^j \in t_k$ and

$$\{\mathbf{B}^j, \varphi_1^j \mathbf{C}_1^j, \ldots, \varphi_k^j \mathbf{C}_k^j\} \vdash_S \bot.$$

(This relies on 5.1(d).) Put $\mathbf{C}_1 = \mathbf{C}_1^1 \wedge \cdots \wedge \mathbf{C}_1^l, \ldots, \mathbf{C}_k = \mathbf{C}_k^1 \wedge \cdots \wedge \mathbf{C}_k^l$. Then for each j we find

$$\vdash_S \mathbf{B}^j \to -(\varphi_1^j \mathbf{C}_1 \wedge \cdots \wedge \varphi_k^j \mathbf{C}_k).$$

Since $\square^{q_j}\mathbf{B}^j \in t'_j$, we have $\square^{q_j} -(\varphi_1^j \mathbf{C}_1 \wedge \cdots \wedge \varphi_k^j \mathbf{C}_k) \in t'_j$, and so given $uR_S^{p_j}t'_j$, $\Diamond^{p_j}\square^{q_j}-(\varphi_1^j \mathbf{C}_1 \wedge \cdots \wedge \varphi_k^j \mathbf{C}_k) \in u$, for each j. But $H' \subseteq u$. It follows that

$$\square^{m_1}\Diamond^{n_1} - \mathbf{C}_1 \vee \cdots \vee \square^{m_k}\Diamond^{n_k} - \mathbf{C}_k \in u,$$

i.e., that $\square^{m_i}\Diamond^{n_i}-\mathbf{C}_i \in u$ for some i. Given $uR_S^{m_i}t_i$, $\Diamond^{n_i}-\mathbf{C}_i \in t_i$. By the definition of $\mathbf{C}_i$, $\Diamond^{n_i}-\mathbf{C}_i^j \in t_i$ for some j, which, given the consistency of t_i, contradicts $\square^{n_i}\mathbf{C}_i^j \in t_i$ for all j. We conclude that for some j, Γ_j is S-consistent. Selecting t''_j a maximal consistent S-extension of Γ_j by 0.1, we have $t'_j R^{q_j} t''_j$ and $t_i R^{\varphi_i^j, n_i} t''_j$ for each i ($1 \leq i \leq k$), by 5.3 and the definition of Γ_j. Thus $\mathcal{K}_S$ satisfies h'.

Corollary. $\vdash_{KH'} \mathbf{A}$ iff $\vDash^{\mathcal{K}} \mathbf{A}$ for all $\mathcal{K}$ satisfying h'.

We have already observed that our completeness result for G' (Corollary 4.2) is only a special case of this new one; so, therefore, are all the completeness results gleaned in the last section. Let us turn, therefore, to what is new.

ℂ:[24] Let us take $k = 1$ in H'. Then the scheme

$$\Diamond^m \square^n \mathbf{A} \to \varphi_1 \mathbf{A} \vee \cdots \vee \varphi_l \mathbf{A},$$

for affirmative modalities $\varphi_1, \ldots, \varphi_l$, may be seen complete with respect to all $\mathcal{K}$ satisfying

$$(\forall u, t)(uR^m t \to tR^{\varphi_1, n}u \vee \cdots \vee tR^{\varphi_l, n}u).$$

---

[24] Case 𝔸 is found on p. 58 f., case 𝔹 on p. 59 f.

Subcases: (i) Put $l = 1$. Then the scheme

$$\Diamond^m \Box^n \mathbf{A} \to \varphi \mathbf{A},$$

for any affirmative φ, is complete with respect to $\mathcal{K}$ satisfying

$$(\forall u, t)(uR^m t \to tR^{\varphi,n} u).$$

(This is the limiting case of H' which is itself a considerable generalization of G'.) We note some special cases of this which are not themselves covered by G'. Thus

$$\Diamond^m \Box^n \mathbf{A} \to \Diamond^p \Box^q \mathbf{A}$$

can be seen complete with respect to $\mathcal{K}$ satisfying

$$(\forall u, t)(uR^m t \to (\exists t_1)(uR^p t_1 \land (\forall t_1')(t_1 R^q t_1' \to tR^n t_1'))).$$

This scheme contraposes into $\Box^p \Diamond^q \mathbf{A} \to \Box^m \Diamond^n \mathbf{A}$. As further special cases, putting $m = 0$ or $n = 0$ or both, we have the schemes

$$\Box^p \Diamond^q \mathbf{A} \to \Diamond^n \mathbf{A},$$
$$\Box^p \Diamond^q \mathbf{A} \to \Box^m \mathbf{A},$$
$$\Box^p \Diamond^q \mathbf{A} \to \mathbf{A},$$

complete with respect to the conditions

$$(\forall u)(\exists t_1)(uR^p t_1 \land (\forall t_1')(t_1 R^q t_1' \to uR^n t_1')),$$
$$(\forall u, t)(uR^m t \to (\exists t_1)(uR^p t_1 \land (\forall t_1')(t_1 R^q t_1' \to t = t_1'))),$$
$$(\forall u)(\exists t_1)(uR^p t_1 \land (\forall t_1')(t_1 R^q t_1' \to u = t_1')).$$

(Other special cases fall under G').

(ii) Notice, first, that our scheme under $\mathbb{C}$ contraposes into

$$\varphi_1' \mathbf{A} \land \cdots \land \varphi_l' \mathbf{A} \to \Box^m \Diamond^n \mathbf{A},$$

where $\varphi_1', \ldots, \varphi_l'$ are the duals of $\varphi_1, \ldots, \varphi_l$. Of the many available special cases, we here consider only

$$\mathbf{A} \land \Diamond^p \Box^q \mathbf{A} \to \Box^m \Diamond^n \mathbf{A},$$

which is in effect $\mathbf{A} \to G'$. This is the case $l = 2$, $\varphi_1' = \phi$, $\varphi_2' = \Diamond^p \Box^q$, so that $\varphi_1 = \phi$, $\varphi_2 = \Box^p \Diamond^q$. This is complete for the condition

$$(\forall u, t)(uR^m t \to tR^{\phi,n} u \land t R^{\Box_p \Diamond_q, n} u).$$

This may easily be transformed into

$$(\forall u, t, t')(uR^m t \land uR^p t' \land {-}tR^n u \to (\exists t_0)(tR^n t_0 \land t' R^q t_0)),$$

which should be compared with g'. Taking $n = 0$ in this, we find that the scheme

$$\mathbf{A} \wedge \Diamond^p \Box^q \mathbf{A} \to \Box^m \mathbf{A}$$

is complete for

$$(\forall u, t, t')(uR^m t \wedge uR^p t' \wedge u \neq t \to t'R^q t),$$

which should be compared with the generalized euclidean condition of Section 4.

𝔻: Take $l = 1$ in H'. Then the scheme

$$\Diamond^{m_1}\Box^{n_1}\mathbf{A}_1 \wedge \cdots \wedge \Diamond^{m_k}\Box^{n_k}\mathbf{A}_k \to \Box^p \Diamond^q (\varphi_1 \mathbf{A}_1 \wedge \cdots \wedge \varphi_k \mathbf{A}_k)$$

for affirmative $\varphi_1, \ldots, \varphi_k$ is complete with respect to $\mathcal{K}$ satisfying

$$(\forall u, t_1, \ldots, t_k)(uR^{m_1} r_1 \wedge \cdots \wedge uR^{m_k} t_k \to (\forall t')(uR^p t' \to$$
$$(\exists t'')(t'R^q t'' \wedge t_1 R^{\varphi_1, n_1} t'' \wedge \cdots \wedge t_k R^{\varphi_k, n_k} t''))).$$

Subcases: (i) Put $n_1, \ldots, n_k = 0, p = 0$; $m_1, \ldots, m_k = 1, q = 1$; $\varphi_1, \ldots, \varphi_k = \Diamond$. The scheme is

$$\Diamond \mathbf{A}_1 \wedge \cdots \wedge \Diamond \mathbf{A}_k \to \Diamond(\Diamond \mathbf{A}_1 \wedge \cdots \wedge \Diamond \mathbf{A}_k).$$

Note that $tR^{\Diamond,0}t'$ iff $t'Rt$. The corresponding condition is

$$(\forall u, t_1, \ldots, t_k)(uRt_1 \wedge \cdots \wedge uRt_k$$
$$\to (\exists t_0)(uRt_0 \wedge t_0 Rt_1 \wedge \cdots \wedge t_0 Rt_k)).$$

By contraposition, this scheme is equivalent to

$$\Box(\Box \mathbf{A}_1 \vee \cdots \vee \Box \mathbf{A}_k) \to \Box \mathbf{A}_1 \vee \cdots \vee \Box \mathbf{A}_k.$$

(ii) A slight variation gives, for the scheme

$$\Box(\Diamond \mathbf{A} \vee \Box \mathbf{A}_1 \vee \cdots \vee \Box \mathbf{A}_k) \to \Diamond \mathbf{A} \vee \Box \mathbf{A}_1 \vee \cdots \vee \Box \mathbf{A}_k,$$

completeness with respect to the condition

$$(\forall u, t_1, \ldots, t_k)(uRt_1 \wedge \cdots \wedge uRt_k$$
$$\to (\exists t_0)(uRt_0 \wedge t_0 Rt_1 \wedge \cdots \wedge t_0 Rt_k \wedge (\forall t')(t_0 Rt' \to uRt'))). \quad (4)$$

(iii) We note some special cases where $k = 2$. The schemes

$$\Diamond \mathbf{A} \wedge \Diamond \mathbf{B} \to \Diamond(\mathbf{A} \wedge \mathbf{B}),$$
$$\Diamond \mathbf{A} \wedge \Diamond \mathbf{B} \to \Diamond(\Diamond \mathbf{A} \wedge \mathbf{B}),$$
$$\Diamond \mathbf{A} \wedge \Diamond \mathbf{B} \to \Diamond(\Diamond \mathbf{A} \wedge \Diamond \mathbf{B}),$$

turn out to be complete respectively for the conditions

$$(\forall u, t, t')(uRt \land uRt' \to t = t'),$$
$$(\forall u, t, t')(uRt \land uRt' \to tRt'),$$
$$(\forall u, t, t')(uRt \land uRt' \to (\exists t_0)(uRt_0 \land t_0Rt \land t_0Rt')).$$

The first of these conditions was just the condition used in connection with the scheme F in Section 4, and the second is just the condition that R is euclidean. Thus the first scheme is deductively equivalent to F and the second to E (in the system K); a direct proof is not hard to find.

We now consider the particular scheme to be called H (after Hintikka):

$$H: \quad \Diamond \mathbf{A} \land \Diamond \mathbf{B} \to \Diamond(\mathbf{A} \land \mathbf{B}) \lor \Diamond(\Diamond \mathbf{A} \land \mathbf{B}) \lor \Diamond(\mathbf{A} \land \Diamond \mathbf{B}),$$

which will prove important in connection with tense logic in the next chapter;[25] indeed, it was generalizing on a completeness result for H that led the authors to consider the scheme H', as a generalization on a result for G led to that for G'. H is the case of H' in which $k = 2$, $l = 3$, $n_1, n_2 = 0$, $p_1, p_2, p_3 = 0$, $m_1, m_2, q_1, q_2, q_3 = 1$, and φ_1^1, $\varphi_2^1 = \phi$, $\varphi_1^2 = \Diamond$, $\varphi_2^2 = \phi$, $\varphi_1^3 = \phi$, $\varphi_2^3 = \Diamond$. The condition h' boils down here (compare $\mathbb{D}$(iii) above) to

$$(h): \quad (\forall u, t, t')(uRt \land uRt' \to t = t' \lor tRt' \lor t'Rt).$$

Thus theoremhood in KH is equivalent to validity in all $\mathcal{K}$ satisfying h. The condition (h) is closely related to the condition that R be *connected*.

A slightly stronger scheme than H is

$$H^+: \quad \Diamond \mathbf{A} \land \Diamond \mathbf{B} \to \Diamond(\Diamond \mathbf{A} \land \mathbf{B}) \lor \Diamond(\mathbf{A} \land \Diamond \mathbf{B}),$$

which is complete for the condition

$$(h^+): \quad (\forall u, t, t')(uRt \land uRt' \to tRt' \lor t'Rt).$$

The system $T4H^+$ is in fact identical with $S4.3$ of Dummett–Lemmon [1], as is shown, e.g., by Prior [4].[26] We may observe that, if R is euclidean, then R satisfies h^+; if R satisfies h^+, then R

---

[25] That chapter was never written.

[26] In Dummett–Lemmon [1] *S4.3* is defined as the smallest normal extension of *S4* that contains all instances of the schema $\Box(\Box \mathbf{A} \to \Box \mathbf{B}) \lor \Box(\Box \mathbf{B} \to \Box \mathbf{A})$. Also noted there is an observation by Geach that the schema $\Box(\Box \mathbf{A} \to \mathbf{B}) \lor \Box(\Box \mathbf{B} \to \mathbf{A})$ would do just as well. Cf. the discussion on p. 80 f. below.

satisfies g and (trivially) R satisfies h. By our completeness results, these remarks justify the inclusions

$$K \subseteq KG \subseteq KH^+ \subseteq KE; \tag{5}$$

$$K \subseteq KH \subseteq KH^+ \subseteq KE. \tag{6}$$

If R is reflexive, R satisfies h iff R satisfies h^+. As additions to T, therefore, H and H^+ are deductively equivalent. In the extra presence of 4 we find:

$$T4 \subseteq T4G \subseteq T4H = T4H^+ \subseteq T5. \tag{7}$$

As already remarked, $T4G$ and $T4H$ are the systems $S4.2$ and $S4.3$, respectively. As usual, it is no difficult matter to show that all these inclusions are proper.

We may also gear our present completeness results to a further set of axioms discussed in the literature. Let us say that a sentence **A** is *fully modalized* iff all occurrences of sentence letters in **A** lie within the scope of an occurrence of $\Box$. If **A** is fully modalized, then the normal form techniques of Section 3 lead to a conjunctive normal form for **A** where each conjunct takes the form

$$\Diamond \mathbf{C} \vee \Box \mathbf{C}_1 \vee \cdots \vee \Box \mathbf{C}_k,$$

for $k \geq 0$. (We may always introduce some **C** since $\vdash_K \Box \mathbf{A} \leftrightarrow \Box \mathbf{A} \vee \Diamond \bot$.) Now consider the set of axioms

N: $\Box \mathbf{A} \to \mathbf{A}$, for fully modalized **A**,

suggested in Lemmon [1] in connection with certain deontic logics, and further discussed in Sobociński [2]. Then adding N is equivalent to adding as axioms *all* sentences of the form

$$\Box(\Diamond \mathbf{A} \vee \Box \mathbf{A}_1 \vee \cdots \vee \Box \mathbf{A}_k) \to \Diamond \mathbf{A} \vee \Box \mathbf{A}_1 \vee \cdots \vee \Box \mathbf{A}_k,$$

for *any* $k \geq 0$, by the above normal form for fully modalized **A**. By $\mathbb{D}$(ii) supra, it follows that theoremhood in KN is equivalent to validity in all $\mathcal{K}$ satisfying (4) above *for every k*.

If N is added to $K4$, a simpler situation emerges. Note first that, if in (4) $k = 0$, the condition becomes simply

$$(\forall u)(\exists t_0)(uRt_0 \wedge (\forall t')(t_0 R t' \to uRt')),$$

which is the correct condition for the scheme $\Box \Diamond \mathbf{A} \to \Diamond \mathbf{A}$ (compare $\mathbb{C}$(i) above). This condition entails that R is serial, so that $\vdash_{KN} \Box \mathbf{A} \to \Diamond \mathbf{A}$, i.e., $D \subseteq KN$. (A direct proof is given in Sobociński.) It now turns out that the system *K4N* can be more simply axiomatized by adding to *D4* the scheme

N_0: $\Diamond \mathbf{A} \wedge \Diamond \mathbf{B} \to \Diamond(\Diamond \mathbf{A} \wedge \Diamond \mathbf{B})$

(equivalently $\Box(\Box \mathbf{A} \vee \Box \mathbf{B}) \to \Box \mathbf{A} \vee \Box \mathbf{B}$) for which a completeness result is given under $\mathbb{D}$(iii) above. This can be shown either directly or by comparing the relevant R-conditions. Thus $K4N = D4N_0$.

There are two further generalizations of our result for H', particular cases of which will be useful in what follows. They will round off our attempts to gain completeness results for normal *K*-systems. We shall consider $\Box^n H'$, the scheme resulting from H' by prefixing $\Box^n$ to the whole, and $\varphi \mathsf{T} \to H'$ (where φ is any affirmative modality).

Let $h'(u)$ be the condition on u which results from h' by dropping the initial universal quantifier $\forall u$. We show completeness of $\Box^n H'$ with respect to all $\mathcal{K}$ satisfying the condition

$$(\forall u, u')(uR^n u' \to h'(u')). \tag{8}$$

Theorem 5.6. $\vDash^{\mathcal{K}} \Box^n H'$ for any $\mathcal{K}$ satisfying (8) above.

Proof is a trivial modification of that of 5.4.

Theorem 5.7. *For any consistent normal K-system S, if S contains* $\Box^n H'$, *then* $\mathcal{K}_S$ *satisfies* (8) *above*.

Proof. We select $u, u' \in U_S$, and suppose $uR^n u'$, i.e., $\{\mathbf{A} : \Box^n \mathbf{A} \in u\} \subseteq u'$. Since $\Box^n H' \in u$, $H' \in u'$. The proof that $h'(u')$ holds for R_S is now identical with that of 5.5, taking u there as u'.

Corollary. $\vdash_{K \Box^n H'} \mathbf{A}$ iff $\vDash^{\mathcal{K}} \mathbf{A}$ for all $\mathcal{K}$ satisfying (8) above.

Second, we show completeness of $\varphi \mathsf{T} \to H'$ (for affirmative φ) with respect to all $\mathcal{K}$ satisfying the condition

$$(\forall u, u')(u' R^{\varphi, 0} u \to h'(u)). \tag{9}$$

Theorem 5.8. $\vDash^{\mathcal{K}} \varphi \mathsf{T} \to H'$ for any $\mathcal{K}$ satisfying (9) above (where φ is affirmative).

Proof. For $\mathcal{K}$ satisfying (9), and $\mathcal{U}$ on $\mathcal{K}$, $u \in U$, suppose $\vDash^{\mathcal{U}}_u \varphi \mathsf{T}$. Where φ by 5.1(a) has the standard form $\Box^{p_1} \Diamond^{q_1} \cdots \Box^{p_k} \Diamond^{q_k}$, this gives

$$(\forall t'_1)(uR^{p_1} t'_1 \to \cdots (\exists t_k)(t'_k R^{q_k} t_k \wedge \vDash^{\mathcal{U}}_{t_k} \mathsf{T}) \cdots).$$

So we have u' such that

$$(\forall t'_1)(uR^{p_1} t'_1 \to \cdots (\exists t_k)(t'_k R^{q_k} t_k \wedge u' = t_k),$$

i.e., $u'R^{\varphi,0}u$ by (1) and (2) above. Hence $h'(u)$ holds by (9). The proof that $\vDash_u^{\mathcal{U}} H'$ now follows that of 5.4.

Theorem 5.9. *For any consistent normal K-system S, if S contains $\varphi\mathsf{T} \to H'$ (for affirmative φ), then $\mathcal{K}_S$ satisfies (9) above.*

Proof. We select $u, u' \in U_S$, and suppose $u'R_S^{\varphi,0}u$. By 5.3, $\{\varphi\mathbf{A} : \mathbf{A} \in u'\} \subseteq u$. Since $\mathsf{T} \in u'$, $\varphi\mathsf{T} \in u$; but $\varphi\mathsf{T} \to H' \in u$; hence $H' \in u$. The proof that $h'(u)$ holds for R_S is now identical with that of 5.5.

Corollary. $\vdash_{K(\varphi\mathsf{T} \to H')} \mathbf{A}$ *iff* $\vDash^{\mathcal{K}} \mathbf{A}$ *for all $\mathcal{K}$ satisfying (9) above (where φ is affirmative).*

The most interesting special case of this corollary is that in which φ is $\Diamond$. Given that $u'R^{\Diamond,0}u$ iff uRu', we find that the scheme

$$\Diamond\mathsf{T} \to H'$$

is complete for the condition

$$(\forall u, u')(uRu' \to h'(u)). \tag{10}$$

The special cases in which we are most interested are the following. The scheme

U: $\Box(\Box\mathbf{A} \to \mathbf{A})$

can be seen complete for the condition

(u): $(\forall u, t)(uRt \to tRt)$

(for H' collapsing via G' into T, i.e., $\Box\mathbf{A} \to \mathbf{A}$, h' collapses into reflexivity; (8) then gives (u).) Similarly

C: $\Box(\Diamond\Box\mathbf{A} \to \mathbf{A})$

is complete for the condition

(c): $(\forall u, t, t')(uRt \wedge tRt' \to t'Rt)$

(compare B and the symmetry condition). Finally, for

L: $\Diamond\mathsf{T} \to (\Box\mathbf{A} \to \mathbf{A})$

we find the completeness condition

(l): $(\forall u, t)(uRt \to uRu)$,

by (10) above. (c) should be compared carefully with (u). Of course, *all* our earlier particular results may be thus generalized via the corollaries to 5.7 and 5.9.

Concerning these new systems KU, KC, KL, notice that $KU \subseteq T$, $KL \subseteq T$ since $\vdash_T U$ by RN, $\vdash_T L$ by PC, and $KC \subseteq KB$ by RN. As systems intermediate between K and T, the systems D, KU, KL are each independent of the others, however. Since if R is euclidean, R satisfies (u), $KU \subseteq KE$; as usual, a direct derivation of U in KE can be found. The scheme U is important for deontic logic, and we shall discuss it further in the next chapter;[27] indeed, the system KU is the system OT of Smiley [1].[28] Notice also that $DL = T$, but $DU \subset T$ (this last results from the fact that we can construct $\mathcal{K}$ whose R is serial and satisfies (u) without being reflexive: for example, $\mathcal{K} = \langle \{u, t\}, \{\langle u, t\rangle, \langle t, t\rangle\}\rangle$). Notice finally that, if R is symmetric, then R satisfies (u) iff R satisfies (l). This means that $KBU = KBL$; a direct proof—not too easy—is left to the reader. It follows that $K5$, which we know to contain both B and E, contains both U and L as well. (**A** is a theorem of $K5$ iff **A** is valid in all $\mathcal{K}$ whose R is an equivalence relation *in its field*, but it does not follow that such an R is *reflexive*; this explains the distinction between $K5$ and $T5$.) The scheme C will come into play in Chapter II.[27]

We conclude this section with some remarks concerning completeness results in general, some concerning limitations on our present techniques for obtaining them, and some designed to relate our new completeness results to those already obtained in Section 2.

First, notice that there is only one inconsistent K-system: its theorems are the class of *all* modal sentences by *PC*, and we have for it a trivial completeness result; theoremhood in it is equivalent to validity in all $\mathcal{K}$ belonging to the null class of world systems $\mathcal{K}$. Accordingly, let Γ be *any non-empty* class of world systems, and let us write $\vDash^\Gamma \mathbf{A}$ to mean $\vDash^\mathcal{K} \mathbf{A}$ for all $\mathcal{K} \in \Gamma$. Put S_Γ for the set of modal sentences **A** such that $\vDash^\Gamma \mathbf{A}$. Then S_Γ will be a consistent normal K-system. For $PC \subseteq S_\Gamma$, $\perp \notin S_\Gamma$ (since Γ is non-empty), $A2 \subseteq S_\Gamma$, if $\mathbf{A}, \mathbf{A} \rightarrow \mathbf{B} \in S_\Gamma$ then $\mathbf{B} \in S_\Gamma$, and if $\mathbf{A} \in S_\Gamma$ then $\Box \mathbf{A} \in S_\Gamma$ by 2.1, Corollary 1. Further, if, for classes of world systems Γ, Γ', we have $\Gamma \subseteq \Gamma'$, then $S_{\Gamma'} \subseteq S_\Gamma$. One (non-constructive) way of determining a consistent normal K-system is, therefore, to specify a non-empty class Γ of world systems. From this standpoint, our completeness results can be re-read as giving, for alternative specifications of Γ by means of various conditions in R, an axiomatic formulation of S_Γ. In this connection, it is worth bearing in mind that not all distinct specifications of Γ lead to distinct systems S_Γ; for example, if we specify Γ as the class of *connected* world systems, even though Γ is a proper subclass of the totality of world systems, S_Γ is simply K, as is

---

[27] That chapter was never written.
[28] In [1] Smiley actually calls his system **OM**.

shown by 1.1, Corollary 2, and 2.9. But, if the R-condition C_1 determining Γ_1 entails the R-condition C_2 determining Γ_2, then $\Gamma_1 \subseteq \Gamma_2$ and so $S_{\Gamma_2} \subseteq S_{\Gamma_1}$. We have appealed to this fact frequently (and tacitly) in our earlier results. These remarks may help to put our present achievements in better perspective.

It cannot, conversely, be said that every consistent normal K-system S determines a class Γ_S of world systems such that $\vdash_S \mathbf{A}$ iff $\vDash^{\Gamma_S} \mathbf{A}$. For example, the system $S = K_{\mathbf{P}_0}$ (K with sole extra axiom $\mathbf{P}_0$) is consistent and normal, yet there can be no class Γ such that $\vdash_S \mathbf{A}$ iff $\vDash^{\Gamma} \mathbf{A}$, if only because it is easy to design $\mathcal{K}$ such that for some $\mathcal{U}$ on $\mathcal{K}$ $\vDash^{\mathcal{U}} \mathbf{A}$ if $\vdash_S \mathbf{A}$, yet for other $\mathcal{U}$ on $\mathcal{K}$ $\vDash_u^{\mathcal{U}} -\mathbf{P}_0$ for $u \in U$. (Compare the remarks following 2.14.) However, it seems reasonable to conjecture that, if a consistent normal K-system S is *closed with respect to substitution instances* (if $\mathbf{A} \in S$, then, for all $\mathbf{A}' \in \Gamma(\mathbf{A})$, $\mathbf{A}' \in S$), then S determines a class Γ_S of world systems such that $\vdash_S \mathbf{A}$ iff $\vDash^{\Gamma_S} \mathbf{A}$. We have no proof of this conjecture. But to prove it would be to make a considerable difference to our theoretical understanding of the general situation.[29]

What limitations are there to our present techniques? The simplest scheme which is not covered by them is

M: $\Box \Diamond \mathbf{A} \to \Diamond \Box \mathbf{A}$.

(In this regard, compare the schemes under $\mathbb{C}$(i) above.) We know of no completeness result of our present kind which covers M. Notice that M is equivalent to the scheme

M': $\Diamond(\Diamond \mathbf{A} \to \Box \mathbf{A})$;

we *do* have a completeness result for $\Diamond \mathbf{A} \to \Box \mathbf{A}$ (part of the scheme F of Section 4), with respect to the condition

$$(\forall u, t, t')(uRt \wedge uRt' \to t = t').$$

By way of a partial result, consider the schemes

M_k: $\Diamond((\Diamond \mathbf{A}_1 \to \Box \mathbf{A}_1) \wedge \cdots \wedge (\Diamond \mathbf{A}_k \to \Box \mathbf{A}_k))$, $(k \geq 1)$.

Then $M_1 = M'$, deductively equivalent to M. Let KM^∞ be the result of adding to K all schemes M_k for any k, and define the condition (m^∞) thus:

(m^∞) $(\forall u)(\exists t_0)(uRt_0 \wedge (\forall t, t')(t_0 Rt \wedge t_0 Rt' \to t = t'))$.

---

[29] This important conjecture has since been disproved. See Kit Fine, 'An incomplete logic containing S4', *Theoria*, vol. 40 (1974), pp. 23–29, and S. K. Thomason, 'An incompleteness theorem in modal logic', *ibid.*, pp. 30–34.

We establish first completeness for KM^∞ with respect to all $\mathcal{K}$ satisfying m^∞. Thus

Theorem 5.10. $\vDash^\mathcal{K} M_k$ *for any k, for all $\mathcal{K}$ satisfying m^∞.*

Proof. Select $\mathcal{K}$ satisfying m^∞, $\mathcal{U}$ on $\mathcal{K}$, $u \in U$. By m^∞, we find t_0 such that $(\forall t, t')(t_0 R t \wedge t_0 R t' \rightarrow t = t')$. It is easily verified that $\vDash^\mathcal{U}_{t_0} \Diamond \mathbf{A}_i \rightarrow \Box \mathbf{A}_i$ for each i ($1 \leq i \leq k$). Hence $\vDash^\mathcal{U}_u M_k$.

Theorem 5.11. *Let S be a consistent normal K-system such that $M_k \subseteq S$ for every k; then $\mathcal{K}_S$ satisfies m^∞.*

Proof. Select $u \in U_S$, and consider

$$\Gamma = \{\mathbf{A} : \Box \mathbf{A} \in u\} \cup \{\Diamond \mathbf{A}_i \rightarrow \Box \mathbf{A}_i : \text{all } i\},$$

where $\mathbf{A}_1, \ldots, \mathbf{A}_n, \ldots$ is some enumeration of *all* modal sentences. Suppose Γ S-inconsistent. Then we can find sentences $\mathbf{B}$, $\mathbf{A}_{i_1}, \ldots, \mathbf{A}_{i_k}$ such that $\Box \mathbf{B} \in u$ and

$$\{\mathbf{B}, \Diamond \mathbf{A}_{i_1} \rightarrow \Box \mathbf{A}_{i_1}, \ldots, \Diamond \mathbf{A}_{i_k} \rightarrow \Box \mathbf{A}_{i_k}\} \vdash_S \bot.$$

By *RM*, we conclude that $-\Diamond((\Diamond \mathbf{A}_{i_1} \rightarrow \Box \mathbf{A}_{i_1}) \wedge \cdots \wedge (\Diamond \mathbf{A}_{i_k} \rightarrow \Box \mathbf{A}_{i_k})) \in u$, which contradicts the consistency of u given $M_k \subseteq u$. Since Γ is therefore S-consistent, we find t_0 by 0.1 such that $u R_S t_0$ and $\Diamond \mathbf{A}_i \rightarrow \Box \mathbf{A}_i \in t_0$ for all i. Now suppose $t_0 R_S t$, $t_0 R_S t'$, and yet $t \neq t'$. We accordingly have a sentence $\mathbf{B}$ such that $\mathbf{B} \in t$, $-\mathbf{B} \in t'$. But then $\Diamond \mathbf{B} \in t_0$ by $t_0 R_S t$, whence $\Box \mathbf{B} \in t_0$, whence $\mathbf{B} \in t'$ by $t R_S t'$, contradicting the consistency of t'. This shows that R_S, i.e., $\mathcal{K}_S$, satisfies m^∞.

Corollary. $\vdash_{KM^\infty} \mathbf{A}$ *iff* $\vDash^\mathcal{K} \mathbf{A}$ *for all $\mathcal{K}$ satisfying m^∞.*

In $K4$, the scheme M (or M') yields M_k for each k. Thus suppose $\vdash_{K4M} \Diamond \mathbf{A}_1$, $\vdash_{K4M} \Diamond \mathbf{A}_2$. Then $\vdash_{K4M} \Box \Diamond \mathbf{A}_1$, $\vdash_{K4M} \Box \Diamond \mathbf{A}_2$ by *RN*, so that $\vdash_{K4M} \Diamond \Box \mathbf{A}_2$ by M. Hence $\vdash_{K4M} \Diamond(\Diamond \mathbf{A}_1 \wedge \Box \mathbf{A}_2)$ by *T16* (Section 3), so that $\vdash_{K4M} \Diamond \Diamond (\mathbf{A}_1 \wedge \mathbf{A}_2)$ by *T16* and *RB*. Hence $\vdash_{K4M} \Diamond (\mathbf{A}_1 \wedge \mathbf{A}_2)$ by *4* contraposed. Generalizing, we find that if $\vdash_{K4M} \Diamond \mathbf{A}_1, \ldots, \vdash_{K4M} \Diamond \mathbf{A}_k$, then $\vdash_{K4M} \Diamond (\mathbf{A}_1 \wedge \cdots \wedge \mathbf{A}_k)$ for any $k \geq 1$. Taking $\mathbf{A}_1$ as $\Diamond \mathbf{A}_1 \rightarrow \Box \mathbf{A}_1, \ldots, \mathbf{A}_k$ as $\Diamond \mathbf{A}_k \rightarrow \Box \mathbf{A}_k$, we find that $\vdash_{K4M} M_k$ for any k. By the above corollary, we have

Theorem 5.12. $\vdash_{K4M} \mathbf{A}$ *iff* $\vDash^\mathcal{K} \mathbf{A}$ *for all transitive $\mathcal{K}$ satisfying m^∞.*

Corollary. $\vdash_{T4M} \mathbf{A}$ *iff* $\vDash^\mathcal{K} \mathbf{A}$ *for all quasi-orderings $\mathcal{K}$ satisfying m^∞.*

The interest in M stems partly from the fact that it is the simplest scheme of the form $\varphi \mathbf{A} \rightarrow \psi \mathbf{A}$ for affirmative modalities φ, ψ which is not amenable to the treatment of the present section, but also from the fact that *T4M* is identical to the system *S4.1* of McKinsey [3] (see Sobociński [2]). Our last corollary gives a completeness result for *T4M*. We do not, however, have any completeness result for the system *KM* itself. It might be hoped that theoremhood in *KM* is equivalent to validity in all $\mathcal{K}$ satisfying m^∞. This hope, however, is shattered by consideration of the following world system, say $\mathcal{K}$:

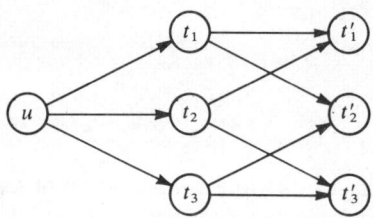

It is easily verified that $\mathcal{K}$, as defined, validates $M = M_1$ but that there exists $\mathcal{U}$ on $\mathcal{K}$ such that an instance of $-M_2$ is verified for $\mathcal{U}$ at u. More complex $\mathcal{K}$ can fairly easily be constructed which validate M_i for all $i \leq k$ but for which we have $\mathcal{U}$ on $\mathcal{K}$ such that an instance of $-M_{k+1}$ is verified for $\mathcal{U}$. Thus M_k does not yield M_{k+1} in K, as it does in *K4*. Accordingly, though we have a completeness result for *KM*$^\infty$ with respect to m^∞, we have none for *KM* itself.[30]

Sobociński [2] considers the addition of G, H^+, and M' (or M) to $T4 (= S4)$ in different combinations.[31] Our results now give completeness for all these possible systems, as they do for the same additions to *K4*. Where inclusions are proper, this is shown easily by considering the relevant completeness conditions: see Prior [4].

How do our present results compare with the completeness results of Section 2? Recall that, in that section (2.17 and 2.18), we showed that $\vdash_{K_{\Gamma(\mathbf{A})}} \mathbf{B}$ iff $\vDash^{\mathcal{U}} \mathbf{B}$ for all $\mathcal{U}$ strongly satisfying $c_\mathbf{A}(u)$, where $\Gamma(\mathbf{A})$ was the set of substitution instances of a sentence $\mathbf{A}$. For example, if $\mathbf{A}$ is the sentence $\Box \mathbf{P}_0 \rightarrow \mathbf{P}_0$, then $K_{\Gamma(\mathbf{A})}$ is simply T, since $\Gamma(\mathbf{A})$ is just the scheme T. We thus, in this case as in others,

---

[30] Since this was written the completeness problem for *KM* has been solved: see Kit Fine, 'Normal forms in modal logic', *Notre Dame journal of formal logic*, vol. 16 (1975), pp. 229–237. For some other interesting results in this connection, see J. F. A. K. van Benthem, 'A note on modal formulae and relational properties', *The journal of symbolic logic*, vol. 40 (1975), pp. 55–58, and 'Modal formulas are either elementary or not $\Sigma\Delta$-elementary', *ibid.*, vol. 41 (1976), pp. 436–438, as well as R. I. Goldblatt, 'First-order definability in modal logic', *ibid.*, vol. 40 (1975), pp. 35–40.

[31] Strictly speaking, what Sobociński discusses in [2] is not H^+ but some schemes related to those mentioned in footnote 26 on p. 69. However, for Lemmon's purposes in this work the difference is not important.

have two distinct results for the system in question. In particular, we know

$$\vdash_T \mathbf{A} \quad \text{iff} \quad \vDash^{\mathcal{K}} \mathbf{A} \text{ for all reflexive } \mathcal{K}, \qquad (11)$$

$$\vdash_T \mathbf{A} \quad \text{iff} \quad \vDash^{\mathcal{U}} \mathbf{A} \text{ for all } \mathcal{U} \text{ strongly satisfying } c_{\square \mathbf{P}_0 \to \mathbf{P}_0}(u). \qquad (12)$$

Now it is easy to see, from the definition of strong satisfaction, that $\mathcal{U} = \langle U, R, \varphi \rangle$ strongly satisfies $c_{\square \mathbf{P}_0 \to \mathbf{P}_0}(u)$ iff

$$(\forall S)(S \in \Sigma_u \to (\forall u)(u \in U \to ((\forall t)(uRt \to t \in S) \to u \in S))). \qquad (13)$$

Putting together (11) and (12), we have

$\vDash^{\mathcal{K}} \mathbf{A}$ for all reflexive $\mathcal{K}$ iff

$$\vDash^{\mathcal{U}} \mathbf{A} \text{ for all } \mathcal{U} \text{ such that (13) holds.} \qquad (14)$$

If we could conclude (14) independently of our present results, the completeness result of this section for T could be derived from that of Section 2. (And so for other cases.) In fact we can show, for any $\mathcal{K} = \langle U, R \rangle$,

$$(\forall u)[(\forall S)(S \subseteq U \to (\forall u)(u \in U \to$$
$$((\forall t(uRt \to t \in S) \to u \in S))) \leftrightarrow uRu]. \qquad (15)$$

(One way is obvious; for the other, pick $S' = \{t : uRt\}$.) In fact, in attempting to infer (11) from (12), we are defeated by the clause $S \in \Sigma_u$ in (13). And other cases are similar. These considerations should point the way towards further research.

In fact, our formulation of the completeness result for H' was guided by the following considerations. We define first what it is for an occurrence of a sentence letter $\mathbf{P}_i$ in $\mathbf{A}$ to be *affirmative* or *negative*: $\mathbf{P}_i$ occurs affirmatively in $\mathbf{P}_i$; any affirmative (negative) occurrence of $\mathbf{P}_i$ in $\mathbf{B}$ is affirmative (negative) in $(\mathbf{C} \to \mathbf{B})$, negative (affirmative) in $(\mathbf{B} \to \mathbf{C})$, and affirmative (negative) in $\square \mathbf{B}$; an occurrence is affirmative or negative only if this follows from the preceding clauses. Now consider a sentence $\Phi = \Phi(\mathbf{P}_1, \ldots, \mathbf{P}_k)$ containing the sentence letters $\mathbf{P}_1, \ldots, \mathbf{P}_k$, and say that Φ is *affirmative* iff all occurrences of $\mathbf{P}_i$ ($1 \leq i \leq k$) in Φ are affirmative. Now $c_\Phi(u)$ will be a condition in R, u containing sentences of the form $t' \in S_i$ ($1 \leq i \leq k$). We abbreviate to

$$c'_\Phi(u)(n_1, \ldots, n_k)$$

the result of replacing in $c_\Phi(u)$ all sentences $t' \in S_i$ by $t_i R^{n_i} t'$ (assuming that the variables t_i do not already occur in $c_\Phi(u)$). Then we *conjecture* that the scheme

$$\Diamond^{m_1}\Box^{n_1}\mathbf{A}_1 \wedge \cdots \wedge \Diamond^{m_k}\Box^{n_k}\mathbf{A}_k \to \Phi(\mathbf{A}_1, \ldots, \mathbf{A}_k),$$

—in which $\Phi(\mathbf{A}_1, \ldots, \mathbf{A}_k)$ represents the result of replacing the atomic sentences $\mathbf{P}_1, \ldots, \mathbf{P}_k$ in Φ by the schematic letters $\mathbf{A}_1, \ldots, \mathbf{A}_k$—is complete for the condition

$$(\forall u, t_1, \ldots, t_k)(uR^{n_1}t_1 \wedge \cdots \wedge uR^{m_k}t_k \to$$
$$c'_\Phi(u)(n_1, \ldots, n_k)),$$

if Φ is affirmative.[32] This holds at least in case $\Phi(\mathbf{A}_1, \ldots, \mathbf{A}_k)$ is the consequent of H' by our completeness result with respect to h'. If the reader verifies this for himself, he will see more clearly the connection between the completeness results in Section 2 and those of the present section. He may also discover some of the snags in the way of developing completeness results for the scheme M.

There remain, therefore, obscurities, as we have seen, in the general situation. We may finally focus attention on one more. For certain conditions c on R (reflexivity, transitivity, symmetry, etc.) we have found schemes Z such that $\vdash_{KZ}\mathbf{A}$ iff $\vDash^{\mathcal{K}}\mathbf{A}$ for all $\mathcal{K}$ whose R satisfies c. But the *general* problem—given a condition c on R to find, if possible, a scheme Z such that $\vdash_{KZ}\mathbf{A}$ iff $\vDash^{\mathcal{K}}\mathbf{A}$ for all $\mathcal{K}$ whose R satisfies c—remains unilluminated: the problem, we may say, of finding a scheme Z to *fit* c. We shall show later that certain conditions on R, e.g., that R is irreflexive, are not in this sense fitted by any scheme Z.[33] And we conjecture that the same is true of other conditions, e.g., that R is asymmetric and that R is antisymmetric. But these general problems remain open problems, as far as this book is concerned.

---

[32] R. I. Goldblatt and Henrik Sahlqvist, independently of one another, have confirmed this conjecture. Sahlqvist's proof was first given in his 'Completeness and correspondence in first and second order semantics for modal logic', unpublished thesis for the cand. real degree at the University of Oslo, 1973, but it is also contained in his paper with the same title in *Proceedings of the Third Scandinavian Logic Symposium*, edited by Stig Kanger, pp. 110–143 (Amsterdam: North-Holland Publishing Co., 1975). Goldblatt's proof is found in his paper 'Solution to a completeness problem of Lemmon and Scott', *Notre Dame journal of formal logic*, vol. 16 (1975), pp. 405–408.

[33] No such proof is given in this book. However, the claim is easily proved, as are the conjectures in the sentence following. Cf. Sahlqvist's works referred to in the previous footnote.

Section 6

We now focus attention more sharply on particular K-systems—in the main, those described in Section 4—and discuss their properties more fully. We attempt in part to do for them what we did for the system K itself in Section 3.

As observed in Section 3, *any* normal K-system provides the rules *RM, RE, RB*, and is classical in the sense of Section 0. We also saw there that K provides the rule of disjunction: whenever $\vdash_K \Box \mathbf{A}_1 \vee \cdots \vee \Box \mathbf{A}_n$ then $\vdash_K \mathbf{A}_i$ for some i ($1 \leq i \leq n$).[34] Unlike the other rules, this rule does not carry over automatically to normal extensions of K—this is reflected partly in the fact that our proof that K provides this rule was nonconstructive, and relied heavily on our completeness result for K (see 3.5 and its proof). For example, the system KU of the last section has as axiom scheme $\Box(\Box \mathbf{A} \to \mathbf{A})$ but it does not have $\Box \mathbf{P}_0 \to \mathbf{P}_0$ as theorem; this suffices to show that KU lacks the rule of disjunction.

We show first that, in addition to K, the systems $D, T, K4, D4, T4$ all do provide this rule.

Theorem 6.1. *For $S = D, T, K4, D4, T4$, if $\vdash_S \Box \mathbf{A}_1 \vee \cdots \vee \Box \mathbf{A}_n$ then $\vdash_S \mathbf{A}_i$ for some i ($1 \leq i \leq n$).*

Proof resembles that of 3.5. For $\mathbf{A}_1, \ldots, \mathbf{A}_n$, suppose not $\vdash_S \mathbf{A}_i$ ($1 \leq i \leq n$); then the sets $S \cup \{-\mathbf{A}_i\}$ are each consistent, yielding $t_i \in U_S$ by 0.1. We construct $\mathcal{K}_S^*$, a safe extension of $\mathcal{K}_S$, as follows. For some $t^* \notin U_S$,

$$\mathcal{K}_S^* = \langle U_S \cup \{t^*\}, R_S \cup \{\langle t^*, t^*\rangle\} \cup \{\langle t^*, t_i\rangle : 1 \leq i \leq n\}$$
$$\cup \{\langle t^*, t'\rangle : t_i R_S t' \text{ for some } i \ (1 \leq i \leq n)\}\rangle.$$

Let us denote the relation just defined by R_S^*. Note that, if R_S is serial, then so is R_S^*; if R_S is reflexive, then so is R_S^*; if R_S is transitive, then so is R_S^*. We define $U_S^* = \langle U_S \cup \{t^*\}, R_S^*, \varphi_S\rangle$. Since $\mathbf{A}_i \notin t_i$, not $\vDash_{t_i}^{\mathcal{U}_S} \mathbf{A}_i$ by 2.8, whence not $\vDash_{t_i}^{\mathcal{U}_S^*} \mathbf{A}_i$ by 1.2. Hence not $\vDash_{t^*}^{\mathcal{U}_S^*} \Box \mathbf{A}_1 \vee \cdots \vee \Box \mathbf{A}_n$. It now follows that not $\vdash_S \Box \mathbf{A}_1 \vee \cdots \vee \Box \mathbf{A}_n$ for $S = D, T, K4, D4, T4$, by the various completeness results for these systems. For example, $\vdash_{D4} \mathbf{A}$ iff $\vDash^{\mathcal{K}} \mathbf{A}$ for all serial transitive

---

[34] Lemmon's manuscript has '$\Box \mathbf{A}_i$'; the necessity symbol has been omitted here. Cf. p. 44.

K, by a combination of 4.4, 4.7. We know that $\mathcal{K}_{D4}$ is both serial and transitive; hence so is $\mathcal{K}^*_{D4}$, by the remarks above.

Corollary. *For $S = D$, T, $K4$, $D4$, $T4$, $\vdash_S \Box \mathbf{A}_1 \vee \cdots \vee \Box \mathbf{A}_n$ iff $\vdash_S \mathbf{A}_i$ for some i $(1 \leq i \leq n)$;*[35] *$\vdash_S \Box \mathbf{A}$ iff $\vdash_S \mathbf{A}$.*

By contrast, addition of the scheme B, $\Diamond \Box \mathbf{A} \to \mathbf{A}$, destroys the rule of disjunction. Thus $\vdash_{KB} \Diamond \Box \Box \mathbf{P}_0 \to \Box \mathbf{P}_0$, whence $\vdash_{KB} \Box \Diamond \Diamond - \mathbf{P}_0 \vee \Box \mathbf{P}_0$, but neither $\vdash_{KB} \Diamond \Diamond - \mathbf{P}_0$ nor $\vdash_{KB} \mathbf{P}_0$. Indeed, as this example shows, any consistent extension of KB which is closed with respect to substitution fails to provide the rule. Further, if we bear in mind that B corresponds to symmetry, we can readily see where attempts to show that KB provides the rule, along the lines of the above proof, fail. We cannot construct R_S^* to be symmetric in case R_S is symmetric without the consequence that $\mathcal{K}_S^*$ is no longer a safe extension of $\mathcal{K}_S$.

Similarly, addition of E, $\Diamond \Box \mathbf{A} \to \Box \mathbf{A}$, or G, $\Diamond \Box \mathbf{A} \to \Box \Diamond \mathbf{A}$, destroys the rule. To see that KH and KH^+ do not provide the rule either, we give first an alternative axiomatization of the two systems, interesting in its own right, from which the fact will emerge obviously. Consider the scheme

$$H_0^+: \quad \Box(\Box \mathbf{A} \to \mathbf{B}) \vee \Box(\Box \mathbf{B} \to \mathbf{A}).$$

We show $KH^+ = KH_0^+$. For suppose $-(\Box(\Box \mathbf{A} \to \mathbf{B}) \vee \Box(\Box \mathbf{B} \to \mathbf{A}))$; then $\Diamond(\Box \mathbf{A} \wedge -\mathbf{B}) \wedge \Diamond(\Box \mathbf{B} \wedge -\mathbf{A})$, whence by H^+

$$\Diamond(\Diamond(\Box \mathbf{A} \wedge -\mathbf{B}) \wedge (\Box \mathbf{B} \wedge -\mathbf{A}))$$
$$\vee \Diamond((\Box \mathbf{A} \wedge -\mathbf{B}) \wedge \Diamond(\Box \mathbf{B} \wedge -\mathbf{A})).$$

Either disjunct leads to absurdity by $T5$. Thus $KH_0^+ \subseteq KH^+$. Conversely, suppose $\Diamond \mathbf{A}$, $\Diamond \mathbf{B}$, yet $-\Diamond(\Diamond \mathbf{A} \wedge \mathbf{B})$, $-\Diamond(\mathbf{A} \wedge \Diamond \mathbf{B})$; then $\Box(\mathbf{A} \to \Box -\mathbf{B}) \wedge \Box(\mathbf{B} \to \Box -\mathbf{A})$. By H_0^+, $\Box(\Box -\mathbf{A} \to -\mathbf{B}) \vee \Box(\Box -\mathbf{B} \to -\mathbf{A})$. But given $\Box(\Box -\mathbf{A} \to -\mathbf{B})$, we have $\Box(\mathbf{B} \to -\mathbf{B})$ and so $\Box -\mathbf{B}$, contradicting $\Diamond \mathbf{B}$; given $\Box(\Box -\mathbf{B} \to -\mathbf{A})$, we have $\Box(\mathbf{A} \to -\mathbf{A})$ and so $\Box -\mathbf{A}$, contradicting $\Diamond \mathbf{A}$. Thus $KH^+ \subseteq KH_0^+$. In a similar, though rather more complex, manner it may be shown that H is deductively equivalent to

$$H_0: \quad \Box(\mathbf{A} \to (\Box \mathbf{B} \to \mathbf{C})) \vee \Box(-\mathbf{A} \to (\Box \mathbf{C} \to \mathbf{B})),$$

i.e., $KH = KH_0$. From these alternative axiom schemes for KH and KH^+, it is obvious at once that neither system provides the rule of disjunction.

---

[35] Lemmon's manuscript has '$\Box \mathbf{A}_i$'; the necessity symbol has been omitted here. Cf. the footnote on p. 79.

SECTION 6

This is a good point at which to remark that the system $T4H = T4H^+ = S4.3$ was originally axiomatized in Dummett–Lemmon [1] by adding to $T4$ the scheme

$$\Box(\Box\mathbf{A} \to \Box\mathbf{B}) \vee \Box(\Box\mathbf{B} \to \Box\mathbf{A}). \tag{1}$$

(1) can easily be seen deductively equivalent to H_0^+ in the presence of T, $\Box\mathbf{A} \to \mathbf{A}$, and 4, $\Box\mathbf{A} \to \Box\Box\mathbf{A}$.

We noted at the beginning of the section that KU fails to provide the rule: similarly, so does KC. By contrast, KN and KL both do provide it. Bear in mind that $\vdash_{KN} \mathbf{A}$ iff $\vDash^{\mathcal{K}} \mathbf{A}$ for all $\mathcal{K}$ satisfying (4) of Section 5 for any n, and note simply that, given $\mathcal{K}_{KN}$ satisfying (4) for any k, it follows that $\mathcal{K}_{KN}^*$ also satisfies (4) for any k, since $t^* R_{KN}^* t^*$. Also $\vdash_{KL} \mathbf{A}$ iff $\vDash^{\mathcal{K}} \mathbf{A}$ for all $\mathcal{K}$ satisfying l, and $\mathcal{K}_{KL}^*$ does satisfy l, since $\mathcal{K}_{KL}$ does. The proof of 6.1, therefore, carries over to these cases. Similarly, the systems $K4N = D4N_0$ and $K4L$ provide the rule too.

Finally, in this regard, we note that the system KM^∞ of the last section provides the rule, since R_S^* can be seen to satisfy m^∞ if R_S does. It follows that $K4M^\infty$ and $T4M^\infty$, which are respectively identical with $K4M$ and $T4M$, also provide the rule. It is worth remarking that all K-systems for which we have established the rule are subsystems of $T4M$. McKinsey–Tarski [1] in effect establish that $T4 = S4$ provides the rule in their Theorem 2.2, and in Theorem 3.10 show that a certain normal extension of $T4$, resulting from adding to $T4$ (in the presence of RN) the scheme

$$\Diamond\Box(\Diamond\mathbf{A} \to \mathbf{A}), \tag{2}$$

also provides the rule. It is not hard to verify directly that this system is just $T4M$ ($S4.1$ of McKinsey [3]), so that their result is equivalent to ours. No stronger system than this appears to be known to provide the rule of disjunction.[36] (For other considerations of this rule, see Lemmon [2] and Kripke [2], pp. 94–95.)

We consider next the problem of decidability for certain systems. In particular, we show how one of the techniques of Section 3, used to show K's decidability, can be extended to yield the decidability of some other K-systems. This is in no sense a *general* technique, however; rather, decidability results are a good deal more elusive than completeness results. In later sections, we shall see that in

---

[36] Let *Grz* be the schema $\Box(\Box(\mathbf{A} \to \Box\mathbf{A}) \to \mathbf{A}) \to \mathbf{A}$. It is known that *T4Grz* is complete with respect to the class of all finite partially ordered world systems (see K. Segerberg, *An essay in classical modal logic*, referred to in n. 20). *T4Grz* is a proper extension of *T4M*, and from the completeness result mentioned it readily follows that *T4Grz* provides the rule of disjunction.

many cases the problem of decidability of one system reduces to that of the decidability of another; this will enable us to enlarge the scope of our results somewhat.[37]

We recall that, for any sentence **A** and set $\Gamma_\mathbf{A}$ of sub-formulas of **A**, we defined an equivalence relation $\equiv$ which partitions U_S into not more than 2^n equivalence classes, where n is the cardinality of $\Gamma_\mathbf{A}$. This led to the definition of

$$\bar{\mathcal{U}}_S = \langle \bar{U}_S, \bar{R}_S, \bar{\varphi}_S \rangle$$

where $\bar{U}_S$ is the set of equivalence classes $[u]$ for $u \in U_S$ and

for $[u], [t] \in \bar{U}_S, [u]\bar{R}_S[t]$

$$\text{iff} \quad (\forall \mathbf{B})(\Box \mathbf{B} \in \Gamma_\mathbf{A} \wedge \Box \mathbf{B} \in u \to \mathbf{B} \in t) \quad (3)$$

$$\bar{\varphi}_S(i) = \{[t] : \mathbf{P}_i \in \Gamma_\mathbf{A} \wedge \mathbf{P}_i \in t\}. \quad (4)$$

We further said that a relation $\bar{\bar{R}}$ on $\bar{U}_S$ is *suitable* iff $uR_S t$ implies $[u]\bar{\bar{R}}[t]$ and $\bar{\bar{R}} \subseteq \bar{R}_S$; also $\bar{\bar{\mathcal{U}}}_S = \langle \bar{\mathcal{U}}_S, \bar{\bar{R}}, \bar{\varphi}_S \rangle$ is suitable if $\bar{\bar{R}}$ is suitable. According to 3.2, if S is a consistent normal K-system and $\bar{\bar{\mathcal{U}}}_S$ suitable, then for any $\mathbf{B} \in \Gamma_\mathbf{A}$, $u \in U_S$,

$$\vDash_{[u]}^{\bar{\bar{\mathcal{U}}}_S} \mathbf{B} \quad \text{iff} \quad \vDash_u^{\mathcal{U}_S} \mathbf{B}. \quad (5)$$

Since $uR_S t$ implies $[u]\bar{R}_S[t]$, $\bar{R}_S$ is always suitable.

It is in the light of this result that we can frequently reduce the decision problem for a normal K-system S to quite manageable proportions, given a completeness result for S. For suppose we know, by completeness, that

$$\vdash_S \mathbf{A} \quad \text{iff} \quad \vDash^{\mathcal{H}} \mathbf{A} \quad \text{for all } \mathcal{H} \text{ satisfying } c,$$

for some condition c. If now we can find suitable $\bar{\bar{\mathcal{H}}}_S$ (for any sentence **A**) which *also* satisfies c, then we have solved the decision problem for S. For select sentence **A**; if $\vdash_S \mathbf{A}$, then $\vDash^{\mathcal{H}} \mathbf{A}$ for all $\mathcal{H}$ satisfying c, so in particular $\vDash^{\bar{\bar{\mathcal{H}}}_S} \mathbf{A}$; on the other hand, if not $\vdash_S \mathbf{A}$, then we have $u \in U_S$ such that not $\vDash_u^{\mathcal{U}_S} \mathbf{B}$ by 2.8, Corollary, so that not $\vDash_{[u]}^{\bar{\bar{\mathcal{U}}}_S} \mathbf{B}$ by (5); thus $-\mathbf{A}$ is satisfiable in $\bar{\bar{\mathcal{H}}}_S$. Since $\bar{\bar{\mathcal{H}}}_S$ is finite, and its size determined by the complexity of **A**, we have a decision procedure for S. Indeed, we can strengthen our completeness result to

$$\vdash_S \mathbf{A} \quad \text{iff} \quad \vDash^{\mathcal{H}} \mathbf{A} \quad \text{for all } \textit{finite } \mathcal{H} \text{ satisfying } c.$$

Now let us take a look at some particular cases. That of T is easiest.

---

[37] These sections were never written; see n. 12.

SECTION 6

Theorem 6.2. *T is decidable.*

Proof. We know $\vdash_T \mathbf{A}$ iff $\vDash^{\mathcal{K}} \mathbf{A}$ for all reflexive $\mathcal{K}$ (4.3). But R_T is itself reflexive; since if $uR_T t$ then $[u]\bar{R}_T[t]$, $\bar{R}_T$ is reflexive too. We know that $\bar{R}_T$ is suitable. Thus $\bar{\mathcal{K}}_T$ is suitable and reflexive, so that the result follows by our preliminary discussion.

Theorem 6.3. *D is decidable.*

Proof. It is easy to see that, since R_D is serial, so is $\bar{R}_D$. The result is immediate by our completeness result for D (4.7).

Theorem 6.4. *K4 is decidable.*

Proof. Since $\vdash_{K4} \mathbf{A}$ iff $\vDash^{\mathcal{K}} \mathbf{A}$ for all transitive $\mathcal{K}$ (4.4), it suffices to define suitable $\bar{\bar{R}}$ which is transitive. For $[u], [t] \in \bar{U}_{K4}$, put

$$[u]\bar{\bar{R}}[t] \quad \text{iff} \quad (\forall \mathbf{B})(\Box \mathbf{B} \in \Gamma_{\mathbf{A}} \wedge \Box \mathbf{B} \in u \to \Box \mathbf{B} \in t \wedge \mathbf{B} \in t). \quad (6)$$

We show first that $\bar{\bar{R}}$ is suitable. Assume $uR_{K4}t$, $\{\mathbf{A} : \Box \mathbf{A} \in u\} \subseteq t$, and select $\mathbf{B}$ such that $\Box \mathbf{B} \in \Gamma_{\mathbf{A}}$, $\Box \mathbf{B} \in u$. Then $\Box\Box \mathbf{B} \in u$, whence $\mathbf{B} \in t, \Box \mathbf{B} \in t$. This gives $[u]\bar{\bar{R}}[t]$. That $\bar{\bar{R}} \subseteq \bar{R}_{K4}$ is immediate. So $\bar{\bar{R}}$ is suitable, and it remains to show that it is transitive. Assume then $[u]\bar{\bar{R}}[t], [t]\bar{\bar{R}}[t']$, and select $\mathbf{B}$ such that $\Box \mathbf{B} \in \Gamma_{\mathbf{A}}$, $\Box \mathbf{B} \in u$. Then $\Box \mathbf{B} \in t$ by $[u]\bar{\bar{R}}[t]$, and $\Box \mathbf{B} \in t'$, $\mathbf{B} \in t'$ follow by $[t]\bar{\bar{R}}[t']$. Hence $[u]\bar{\bar{R}}[t']$, and our proof is complete.

Theorem 6.5. *KB is decidable.*

Proof. Since $\vdash_{KB} \mathbf{A}$ iff $\vDash^{\mathcal{K}} \mathbf{A}$ for all symmetric $\mathcal{K}$ (4.5), it suffices to define suitable $\bar{\bar{R}}$ which is symmetric. For $[u], [t] \in \bar{U}_{KB}$, put

$$[u]\bar{\bar{R}}[t] \quad \text{iff} \quad (\forall \mathbf{B})(\Box \mathbf{B} \in \Gamma_{\mathbf{A}}$$
$$\to (\Box \mathbf{B} \in u \to \mathbf{B} \in t) \wedge (\Box \mathbf{B} \in t \to \mathbf{B} \in u)). \quad (7)$$

Now assume $uR_{KB}t$, $\Box \mathbf{B} \in \Gamma_{\mathbf{A}}$. If $\Box \mathbf{B} \in u$, then $\mathbf{B} \in t$ by $uR_{KB}t$. If $\Box \mathbf{B} \in t$, then $\Diamond \Box \mathbf{B} \in u$ by $uR_{KB}t$, whence $\mathbf{B} \in u$ by the scheme B. Thus $[u]\bar{\bar{R}}[t]$. Since obviously $\bar{\bar{R}} \subseteq \bar{R}_{KB}$, $\bar{\bar{R}}$ is suitable. It is obvious on inspection that $\bar{\bar{R}}$ is symmetric.

Theorem 6.6. *K5 is decidable.*

Proof. We know (a combination of 4.4 and 4.5) that $\vdash_{K5} \mathbf{A}$ iff $\vDash^{\mathcal{K}} \mathbf{A}$ for all transitive symmetric $\mathcal{K}$. It suffices, therefore, to define suitable $\bar{\bar{R}}$ which is both transitive and symmetric. For $[u], [t] \in U_{K5}$, put

$$[u]\bar{\bar{R}}[t] \quad \text{iff} \quad (\forall \mathbf{B})(\Box \mathbf{B} \in \Gamma_{\mathbf{A}}$$
$$\to (\Box \mathbf{B} \in u \to \Box \mathbf{B} \in t \wedge \mathbf{B} \in t) \wedge (\Box \mathbf{B} \in t \to \Box \mathbf{B} \in u \wedge \mathbf{B} \in u)). \quad (8)$$

It is left to the reader to verify that $\bar{\bar{R}}$ is suitable, symmetric, and transitive.

Theorem 6.7. *The systems T4, TB, and T5 are decidable.*

Proof. Use the $\bar{\bar{R}}$ as defined in (6), (7), and (8), respectively, and bear in mind the completeness results in 4.8. It is easily verified that, in the new contexts, $\bar{\bar{R}}$ is reflexive.

Theorem 6.8. *The systems D4, DB are decidable.*

Proof. Use the $\bar{\bar{R}}$ of (6) and (7) respectively; $\bar{\bar{R}}$ is obviously serial. (We do not treat D5, since D5 = T5, as observed in Section 4.)

Of the major systems mentioned in Section 4, this shows the decidability of all but KE, KE4, DE, DE4. We can find no proof, using our present techniques, of the decidability of these systems.[38] The technique does apply to some other systems, however, as we go on to show.

Theorem 6.9. *The systems KU and DU are decidable.*

Proof. We know that $\vdash_{KU} \mathbf{A}$ iff $\vDash^{\mathcal{K}} \mathbf{A}$ for all $\mathcal{K}$ satisfying

$$(u): \quad (\forall u, t)(uRt \to tRt)$$

(Section 5). It suffices, therefore, to define suitable $\bar{\bar{R}}$ satisfying (u). We define, for $[u], [t] \in U_{KU}$,

$$[u]\bar{\bar{R}}[t] \quad \text{iff} \quad (\forall \mathbf{B})(\square \mathbf{B} \in \Gamma_{\mathbf{A}}$$
$$\to (\square \mathbf{B} \in u \to \mathbf{B} \in t) \wedge (\square \mathbf{B} \in t \to \mathbf{B} \in t)). \quad (9)$$

To show $\bar{\bar{R}}$ suitable, assume $uR_{KU}t$. Since $U, \square(\square \mathbf{A} \to \mathbf{A}) \in u$, we have $\square \mathbf{A} \to \mathbf{A} \in t$. This shows that $[u]\bar{\bar{R}}[t]$. Trivially, $\bar{\bar{R}} \subseteq \bar{R}_{KU}$. That $\bar{\bar{R}}$ satisfies (u) is obvious on inspection. The case of DU is left to the reader.

Theorem 6.10. *The system KL is decidable.*

Proof. We know that $\vdash_{KL} \mathbf{A}$ iff $\vDash^{\mathcal{K}} \mathbf{A}$ for all $\mathcal{K}$ satisfying

$$(l): \quad (\forall u, t)(uRt \to uRu)$$

(Section 5). For $[u], [t] \in U_{KL}$, define

$$[u]\bar{\bar{R}}[t] \quad \text{iff} \quad (\forall \mathbf{B})(\square \mathbf{B} \in \Gamma_{\mathbf{A}}$$
$$\to (\square \mathbf{B} \in u \to \mathbf{B} \in t) \wedge (\square \mathbf{B} \in u \to \mathbf{B} \in u)). \quad (10)$$

---

[38] Such proofs exist now: see K. Segerberg, 'Decidability of four modal logics', *Theoria*, vol. 34 (1968), pp. 21–25.

To show $\bar{\bar{R}}$ suitable, assume $uR_{KL}t$. Since $\top \in t$, $\Diamond\top \in u$; since $\Diamond\top \to (\Box\mathbf{A} \to \mathbf{A}) \in u$ by L, $\Box\mathbf{A} \to \mathbf{A} \in u$. This gives $[u]\bar{\bar{R}}[t]$. It follows that $\bar{\bar{R}}$ is suitable, and it clearly satisfies (*l*). (Note that *DL* is just *T*.)

Theorem 6.11. *The systems KU4, KL4, and KBU = KBL are decidable, as is DU4.*

Proof. For each of the first three systems, we merely define suitable $\bar{\bar{R}}$; remaining details are left to the reader. They read

$$[u]\bar{\bar{R}}[t] \quad \text{iff} \quad (\forall \mathbf{B})(\Box\mathbf{B} \in \Gamma_\mathbf{A}$$
$$\to (\Box\mathbf{B} \in u \to \Box\mathbf{B} \in t \wedge \mathbf{B} \in t) \wedge (\Box\mathbf{B} \in t \to \mathbf{B} \in t)), \quad (11)$$

$$[u]\bar{\bar{R}}[t] \quad \text{iff} \quad (\forall \mathbf{B})(\Box\mathbf{B} \in \Gamma_\mathbf{A}$$
$$\to (\Box\mathbf{B} \in u \to \Box\mathbf{B} \in t \wedge \mathbf{B} \in t) \wedge (\Box\mathbf{B} \in u \to \mathbf{B} \in u)), \quad (12)$$

$$[u]\bar{\bar{R}}[t] \quad \text{iff} \quad (\forall \mathbf{B})(\Box\mathbf{B} \in \Gamma_\mathbf{A}$$
$$\to (\Box\mathbf{B} \in u \to \mathbf{B} \in t) \wedge (\Box\mathbf{B} \in t \to \mathbf{B} \in u)$$
$$\wedge (\Box\mathbf{B} \in u \to \mathbf{B} \in u) \wedge (\Box\mathbf{B} \in t \to \mathbf{B} \in t)). \quad (13)$$

(The case of *DU4* is left to the reader; we do not treat *KU5* or *KL5*, since both are identical with *K5*.)

That *T5* is decidable is a result of long standing. The decidability of *T4* appears first due to McKinsey [2], that of *T* to Anderson [1].[39] The first proof of the decidability of *TB* would be seen to be Kripke [2], and that of *D* Lemmon [3]. Since *KU* is Smiley's *OT*, its decidability was first shown in Smiley [1], as was that of *KU4*.[40] The result for *K5* is also in Smiley, though he does not observe the redundancy of the scheme *U* in the presence of *B* and *4*. Other results given here would appear to be new. Bull [1] also shows the decidability of *S4.2* and *S4.3* (our *T4G* and *T4H*). It seems to be an open problem whether *S4.1* (our *T4M*) is decidable or not.[41]

---

[39] It is customary to attribute the result that *T* is decidable to Georg Henrik von Wright; Anderson himself refers to von Wright [2] in his 1954 paper. However, von Wright has published no proof that his proposed decision procedure for *T* actually works. Such a proof was given by Anderson with respect to his own decision procedure, which is also simpler than that of von Wright.

Anderson and Lemmon, and indeed most authors, seem unaware of Ridder's earlier proof that *T* is decidable; see J. Ridder, 'Über modale Aussagenlogiken und ihren Zusammenhang mit Strukturen II', *Indagationes mathematicae*, vol. 14 (1952), pp. 459–467.

[40] See n. 28.

[41] Affirmative solutions to this problem have since been given in R. A. Bull, 'On the extension of S4 with *CLMpMLp*', *Notre Dame journal of formal logic*, vol. 8 (1967), pp. 325–329, and K. Segerberg, 'Decidability of S4.1', *Theoria*, vol. 34 (1968), pp. 7–20.

We devote the remainder of this section to some rather miscellaneous remarks concerning particular modal systems. Many more results than are here given are known; the interested reader should consult Prior [2], [3], and Feys [2] (which has a good bibliography).[42] Since our primary aim is the suitable deployment of theoretical semantics, they lie rather outside the scope of this book.

A result that may help towards a deeper understanding of the system D is the following:

Theorem 6.12. $\vdash_D \mathbf{A}$ iff $\vdash_D \Diamond \mathbf{A}$.

Proof. Given $\vdash_D \mathbf{A}$, $\vdash_D \Box \mathbf{A}$ by RN, whence $\vdash_D \Diamond \mathbf{A}$ by D, $\Box \mathbf{A} \to \Diamond \mathbf{A}$. The converse is harder. Consider $\mathcal{K}_D = \langle U_D, R_D \rangle$, and, for each $u \in U_D$, select $u^* \notin U_L$ such that if $u \ne t$ then $u^* \ne t^*$. We define $\mathcal{K}_D^* = \langle U_D^*, R_D^* \rangle$ where

$$U_D^* = U_D \cup \{u^* : u \in U_D\},$$
$$R_D^* = R_D \cup \{\langle u^*, u \rangle : u \in U_D\}.$$

It follows that $\mathcal{K}_D^*$ is a safe extension of $\mathcal{K}_D$ (compare 1.2). Further, since R_D is serial, so is R_D^*; hence if $\vdash_D \mathbf{A}$ then $\vDash^{\mathcal{K}_D^*} \mathbf{A}$ (4.7). Now suppose $\vdash_D \Diamond \mathbf{A}$. Then for any u^* such that $u \in U_D$ and any $\mathcal{U}$ on $\mathcal{K}_D^*$, $\vDash_{u^*}^{\mathcal{U}} \Diamond \mathbf{A}$. By the construction of R_D^*, since $(\forall t)(u^* R_D^* t \to t = u)$, for any $u \in U_D \vDash_u^{\mathcal{U}} \mathbf{A}$. Taking $\mathcal{U}$ in particular as $\mathcal{U}_D$, for any $u \in U_D \vDash_u^{\mathcal{U}_D} \mathbf{A}$. That $\vdash_D \mathbf{A}$ now follows by 2.8, Corollary.

Corollary. $\vdash_D \mathbf{A}$ iff $\vdash_D \Diamond^m \mathbf{A}$ iff $\vdash_D \Box^n \mathbf{A}$, for any m, n.

This striking property seems fairly special to D.[43] For example, K itself has *no* theorems of the form $\Diamond \mathbf{A}$ (since $\vdash_K \Diamond \mathbf{A} \to \Diamond \top$ by RB, if it had, then $\Diamond \top$ would be a K-theorem, which we know not to be the case). And in the case of T we have $\vdash_T \Diamond(\mathbf{A} \to \Box \mathbf{A})$ but not $\vdash_T \mathbf{P}_0 \to \Box \mathbf{P}_0$.

We saw already in Section 4 that D may alternatively be axiomatized by adding to K the simple sentence $\Diamond \top$ (or indeed $\Diamond^n \top$ for any $n \geq 1$). Trivial alternative axiomatizations of systems with the schemes T, B, 4, or E may be obtained by using their contra-

---

[42] Useful bibliographies are also found in G. E. Hughes and M. J. Cresswell, *An introduction to modal logic* (London: Methuen, 1968), A. N. Prior, *Papers on time and tense* (Oxford: At the Clarendon Press, 1968), and J. Jay Zeman, *Modal logic: The Lewis–modal systems* (Oxford: At the Clarendon Press, 1973).

[43] *KF* has it (see p. 60 f.), and also indefinitely many other systems that are trivial in the sense that they possess finite characteristic matrices.

posed forms, namely:

T_C: $\mathbf{A} \to \Diamond \mathbf{A}$,

B_C: $\mathbf{A} \to \Box \Diamond \mathbf{A}$,

4_C: $\Diamond \Diamond \mathbf{A} \to \Diamond \mathbf{A}$,

E_C: $\Diamond \mathbf{A} \to \Box \Diamond \mathbf{A}$.

These versions have been widely used in the literature. The alternatives H_0 and H_0^+ to H and H^+ are more interesting. Similarly, as deductive equivalents to B (in K) the schemes

$$\Box(\Diamond \mathbf{A} \to \mathbf{B}) \to (\mathbf{A} \to \Box \mathbf{B}),$$

$$\Box(\mathbf{A} \to \Box \mathbf{B}) \to (\Diamond \mathbf{A} \to \mathbf{B})$$

are worth noting. In view of them, we have for KB (as well as DB, TB) the result: $\vdash_{KB} \Diamond \mathbf{A} \to \mathbf{B}$ iff $\vdash_{KB} \mathbf{A} \to \Box \mathbf{B}$. Some alternative axiomatizations of $T5$ were mentioned in Section 4; there are many others in the literature (see Prior [2], and bibliography).[44]

McKinsey's $S4.1$ has a habit of cropping up in unexpected places in the literature. It was originally formulated (McKinsey [3]) by adding to $S4 (= T4)$ the scheme

$$\Box \Diamond \mathbf{A} \wedge \Box \Diamond \mathbf{B} \to \Diamond (\mathbf{A} \wedge \mathbf{B}).$$

This scheme can fairly readily be shown deductively equivalent in $T4$ to the schemes M, M' of Section 5, as well as to the scheme (2) above. An alternative equivalent scheme is

$$-\Box(\Diamond \mathbf{A} \wedge \Diamond - \mathbf{A}),$$

which appears in Dummett–Lemmon [1], p. 257. We may note that, whilst $\vdash_T \Diamond(\Diamond \mathbf{A} \to \mathbf{A})$, $\vdash_T \Diamond(\mathbf{A} \to \Box \mathbf{A})$, the scheme M', $\Diamond(\Diamond \mathbf{A} \to \Box \mathbf{A})$, which is characteristic of $S4.1$, is *not* a theorem scheme even of $T5$ ($S5$).

The situation here is clarified somewhat by discussing the question of the number of distinct *non-equivalent irreducible* modalities available in different systems, a question much aired in the literature. Let us define the *rank* of modality φ as the number of occurrences of the symbols $\Box$ or $\Diamond$ in φ. We say that a modality φ of

---

[44] It is possible that Lemmon had planned to write a bibliography containing also works not specifically referred to in the text. Such a bibliography might have included such works as Alan Ross Anderson, 'Independent axiom schemata for S5', *The journal of symbolic logic*, vol. 21 (1956), pp. 255–256, E. J. Lemmon, 'Alternative postulate sets for S5', *ibid.*, pp. 347–349, and Bolesław Sobociński, 'A contribution to the axiomatization of Lewis's System S5', *Notre Dame journal of formal logic*, vol. 3 (1962), pp. 59–60.

rank m is *reducible to* ψ in system S iff $\vdash_S \varphi A \leftrightarrow \psi A$, where ψ is a modality of rank less than m; if there is no ψ such that φ is reducible to ψ in S, then φ is *irreducible* in S. For example, in *T4* we find $\vdash_{T4} \Box^n A \leftrightarrow \Box A$ for $n \geqq 1$, in virtue of T and 4; hence all modalities $\Box^n$ for $n \geqq 1$ are reducible in *T4* to $\Box$; it can fairly readily be seen that $\Box$ is itself irreducible in *T4*. Modalities φ and ψ are *equivalent* in S iff $\vdash_S \varphi A \leftrightarrow \psi A$; otherwise *non-equivalent*. In the light of these definitions, for given S the question concerning the number of distinct non-equivalent irreducible modalities in S should take on meaning.

It is well known that *T4* provides 14 such (Parry [2], originally). Since the pattern of negative modalities reduplicates that of the affirmative ones, we concentrate on the latter. We need to observe first that the schemes

$$\Box^n A \leftrightarrow \Box A, \tag{14}$$

$$\Diamond^n A \leftrightarrow \Diamond A, \tag{15}$$

$$\Diamond \Box \Diamond \Box A \leftrightarrow \Diamond \Box A, \tag{16}$$

$$\Box \Diamond \Box \Diamond A \leftrightarrow \Box \Diamond A \tag{17}$$

are all theorem schemes of *T4*. (14) is obvious, and (15) follows by contraposition. For (16), note that $\vdash_{T4} \Box \Diamond \Box A \to \Diamond \Box A$ by T, so that $\vdash_{T4} \Diamond \Box \Diamond \Box A \to \Diamond \Diamond \Box A$ by *RB*, hence $\vdash_{T4} \Diamond \Box \Diamond \Box A \to \Diamond \Box A$ by 4; conversely, $\vdash_{T4} \Box A \to \Diamond \Box A$ by T_C, whence $\vdash_{T4} \Box \Box A \to \Box \Diamond \Box A$ by *RM* and $\vdash_{T4} \Box A \to \Box \Diamond \Box A$ by L; thus $\vdash_{T4} \Diamond \Box A \to \Diamond \Box \Diamond \Box A$ by *RB*. (17) follows from (16) by contraposition. Given these results and 5.1(a), any affirmative modality φ is equivalent in *T4* to either the null modality or one of the modalities $\Box$, $\Diamond$, $\Box \Diamond$, $\Diamond \Box$, $\Box \Diamond \Box$, $\Diamond \Box \Diamond$. That no further reductions are available in *T4* can be shown by elementary world systems $\mathcal{K}$. These seven modalities may be displayed in the chart

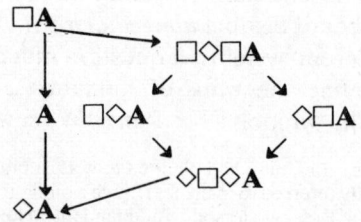

where $\to$ represents a theorem scheme of *T4*. There are seven corresponding negative modalities, making 14 in all.

By adding E, $\Diamond\Box A \to \Box A$, the system $T5$ collapses these 14 down into six. The affirmative modalities may be represented by the chart

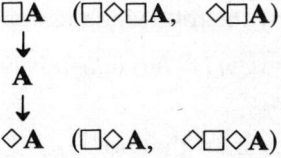

and the corresponding three negative ones are $-$, $\Box-$, and $\Diamond-$. By contrast, the system $T4G$ ($=S4.2$) yields 10 non-equivalent irreducible modalities in view of the scheme $\Diamond\Box A \to \Box\Diamond A$. For $\vdash_{T4G} \Diamond\Box A \to \Box\Diamond\Box A$, and the affirmative modalities of $T4$ collapse into:

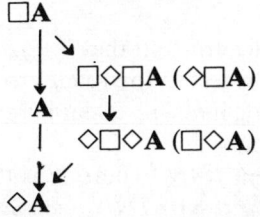

On the other hand, $T4M$ ($=S4.1$) also yields 10 non-equivalent irreducible modalities in view of M, $\Box\Diamond A \to \Diamond\Box A$, since $\vdash_{T4M} \Diamond\Box\Diamond A \to \Diamond\Box A$. This gives, for the affirmative modalities,

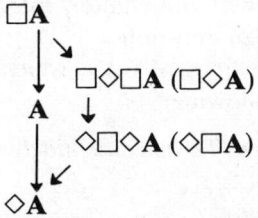

Thus $T4G$ and $T4M$ have exactly 10 distinct non-equivalent irreducible modalities, but a different 10. See further Parry [2], Dummett–Lemmon [1], Sobociński [ ], Prior [ ].[45]

Attention has also been paid to the matter of showing that certain systems had *indefinitely many* distinct non-equivalent irreducible modalities. After Parry [2] showed that Lewis' $S3$ had exactly 42 such modalities, McKinsey [1] proved that $S2$ had indefinitely many. This result was extended to T by Sobociński (see Prior [3]),

---

[45] The editor has not been able to identify the references to Sobociński and Prior.

and, more recently, to $T4^n$ for $n \geq 2$ by Sugihara [1]—a striking result, where 4^n is $\Box^n \mathbf{A} \to \Box^{n+1} \mathbf{A}$ (Section 4). This is one direction in which we can come close to $T4$ without a modalities collapse; another is provided by the following theorem.

Theorem 6.13. *The system D4 has indefinitely many non-equivalent irreducible modalities.*

Proof. It will suffice to show that not $\vdash_{D4} \Box^n \mathbf{P}_0 \to \Box^m \mathbf{P}_0$ for $n > m$. For it will then follow that the modalities $\Box, \Box^2, \Box^3, \ldots$ are all non-equivalent and irreducible ($\Box^n$ cannot reduce to any *other* modality than $\Box^m$ for $m < n$, in view of the containment of $D4$ in $T4$). Consider

$$\mathcal{K} = \langle \{t_0, t_1, \ldots, t_n, \ldots\}, \{\langle t_i, t_j \rangle : i < j\}\rangle.$$

Then $\mathcal{K}$ is transitive and serial, so that if $\vdash_{D4} \mathbf{A}$ then $\vDash^{\mathcal{K}} \mathbf{A}$ by our completeness results. It is left to the reader to design $\mathcal{U}$ on $\mathcal{K}$ such that not $\vDash^{\mathcal{U}}_{t_0} \Box^n \mathbf{P}_0 \to \Box^m \mathbf{P}_0$, if $n > m$. (Bear in mind the completeness result under A(i), Section 4.)

By contrast, the system $K4N$ (where N is the class of sentences $\Box \mathbf{A} \to \mathbf{A}$ for fully modalized $\mathbf{A}$) = $D4N_0$ (see Section 5) provides the same reduction theorems for modalities as $T4$ itself: this was the intention behind N, when originally introduced in Lemmon [1]. Somewhere between $D4$ and $D4N_0$ there is a line to be drawn between systems with indefinitely many non-equivalent irreducible modalities and systems with only finitely many; where exactly this line goes remains an open question.

A final result on the lines of 6.13, which strengthens that of Sobociński [1], is the following:

Theorem 6.14. *The system TB has indefinitely many non-equivalent irreducible modalities.*

Proof. Note that $\vdash_{TB} \mathbf{A}$ iff $\vDash^{\mathcal{K}} \mathbf{A}$ for all reflexive symmetric $\mathcal{K}$. Consider

$$\mathcal{K} = \langle \{\ldots, t_{-n}, \ldots, t_0, \ldots, t_n, \ldots\}, \{\langle t_i, t_j \rangle : i = j \vee i = j+1 \vee$$
$$j = i+1\}\rangle.$$

Then $\mathcal{K}$ is reflexive and symmetric. It is again left to the reader to show that there exists $\mathcal{U}$ on $\mathcal{K}$ such that not $\vDash^{\mathcal{U}}_{t_0} \Box^n \mathbf{P}_0 \to \Box^m \mathbf{P}_0$, if $m > n$. It follows that the modalities $\Box, \Box^2, \Box^3, \ldots$ are all non-equivalent in TB.

BIBLIOGRAPHY

ANDERSON, ALAN ROSS
[1] *Improved decision procedures for Lewis's S4 and von Wright's* M. **The journal of symbolic logic**, vol. 19 (1954), pp. 201–214.

BARCAN, RUTH (MRS J. A. MARCUS)
[1] *A functional calculus of first order based on strict implication.* **The journal of symbolic logic**, vol. 11 (1946), pp. 1–16.

BECKER, ALBRECHT
[1] *Die Aristotelische Theorie der Möglichkeitsschlüsse. Eine logisch-philologische Untersuchung der Kapitel 13–22 von Aristoteles'* **Analytica priora** I. Dissertation, Münster, 1932. Berlin: 1933.

BOCHEŃSKI, I. M.
[1] *A history of formal logic.* Notre Dame, Ind.: University of Notre Dame Press, 1961.

BULL, R. A.
[1] *A note on the modal calculi S4.2 and S4.3.* **Zeitschrift für mathematische Logik und Grundlagen der Mathematik**, vol. 10 (1964), pp. 53–55.

CARNAP, RUDOLF
[1] *Modalities and quantification.* **The journal of symbolic logic**, vol. 11 (1946), pp. 33–64.
[2] *Meaning and necessity: A study in semantics and modal logic.* Chicago: The University of Chicago Press, 1947.

CHISHOLM, RODERICK M.
[1] *Sentences about believing.* **Proceedings of the Aristotelian Society**, vol. 56 (1955–56), pp. 135–148.
[2] *Perceiving: A philosophical study.* Ithaca, N.Y.: Cornell University Press, 1957.

CHURCH, ALONZO
[1] *A formulation of the logic of sense and denotation.* In **Structure, method and meaning: Essays in honor of H. M. Sheffer**, edited by P. Henle, H. M. Kallen and S. K. Langer, pp. 3–24. New York: Liberal Arts Press, 1951.
[2] *Introduction to mathematical logic.* Vol. 1. Princeton: Princeton University Press, 1956.

DUMMETT, M. A. E. & LEMMON, E. J.
[1] *Modal logics between S4 and S5.* **Zeitschrift für mathematische Logik und Grundlagen der Mathematik**, vol. 5 (1959), pp. 250–264.

FEYS, ROBERT
[1] *Les logiques nouvelles des modalités.* **Revue néoscholastique de philosophie**, vol. 40 (1937), pp. 517–553, and vol. 41 (1938), pp. 217–252.
[2] *Modal logics.* Edited with some complements by Joseph Dopp. Louvain: E. Nauwelaerts, and Paris: Gauthiers-Villars, 1965.

FITCH, F. B.
[1] *Symbolic logic: An introduction.* New York: Ronald Press Co., 1952.

GÖDEL, KURT
[1] *Eine Interpretation des intuitionistischen Aussagenkalküls.* **Ergebnisse eines mathematischen Kolloquiums**, vol. 4 (1931–32), pp. 39–40.

HINTIKKA, JAAKKO
[1] *Knowledge and belief: An introduction to the two notions.* Ithaca, N.Y.: Cornell University Press, 1962.
[2] *The modes of modality.* **Proceedings of a colloquium on modal and many-valued logics, Helsinki, 23–26 August, 1962. Acta philosophica Fennica**, fasc. 16 (1963), pp. 65–81.

KAPLAN, DAVID
[1] Review of Kripke [2]. **The journal of symbolic logic**, vol. 31 (1966), pp. 120–122.

KNEALE, WILLIAM & KNEALE, MARY
[1] *The development of logic.* Oxford: At the Clarendon Press, 1962.

KRIPKE, SAUL A.
[1] *A completeness theorem in modal logic.* **The journal of symbolic logic**, vol. 24 (1959), pp. 1–14.
[2] *Semantical analysis of modal logic I: Normal propositional calculi.* **Zeitschrift für Logik und Grundlagen der Mathematik**, vol. 9 (1963), pp. 67–96.

LEMMON, E. J.
[1] *New foundations for Lewis modal systems.* **The journal of symbolic logic**, vol. 22 (1957), pp. 176–186.
[2] *An extension algebra and the modal system T.* **Notre Dame journal of formal logic**, vol. 1 (1960), pp. 3–12.
[3] *Algebraic semantics for modal logics.* **The journal of symbolic logic**, vol. 31 (1966), pp. 46–65, 191–218.

LEWIS, C. I. & LANGFORD, C. H.
[1] *Symbolic logic.* New York and London: The Century Co., 1932.

LUKASIEWICZ, JAN
[1] *Aristotle's syllogistic from the standpoint of modern formal logic.* Second edition. Oxford: At the Clarendon Press, 1957.

McCall, Storrs
[1] *Aristotle's modal syllogisms*. Amsterdam: North-Holland Publishing Co., 1963.

McKinsey, J. C. C.
[1] *Proof that there are infinitely many modalities in Lewis's system S2*. **The journal of symbolic logic**, vol. 5 (1940), pp. 110–112.
[2] *A solution of the decision problem for the Lewis systems S2 and S4 with an application to topology*. **The journal of symbolic logic**, vol. 6 (1941), pp. 117–134.
[3] *On the syntactical construction of systems of modal logic*. **The journal of symbolic logic**, vol. 10 (1945), pp. 83–96.

McKinsey, J. C. C. & Tarski, A.
[1] *Some theorems about the sentential calculi of Lewis and Heyting*. **The journal of symbolic logic**, vol. 13 (1948), pp. 1–15.

Mates, Benson
[1] *Stoic logic*. Berkeley and Los Angeles: University of California Press, 1953.

Ohnishi, M. & Matsumoto, K.
[1] *Gentzen method in modal calculi*. **Osaka mathematical journal**, vol. 9 (1957), pp. 113–130, and vol. 11 (1959), pp. 115–120.

Parry, William Tuthill
[1] *The postulates for strict implication*. **Mind**, n.s. vol. 43 (1934), pp. 78–80.
[2] *Modalities in the Survey system of strict implication*. **The journal of symbolic logic**, vol. 4 (1939), pp. 137–154.

Prior, Arthur
[1] *Diodorean modalities*. **The philosophical quarterly**, vol. 5 (1955), pp. 205–213.
[2] *Formal logic*. Oxford: At the Clarendon Press, 1955. Second edition 1962.
[3] ***Time and modality: Being the John Locke Lectures for 1955–6 delivered in the University of Oxford***. Oxford: At the Clarendon Press, 1955. Second edition 1962.
[4] *K1, K2 and related modal systems*. **Notre Dame journal of formal logic**, vol. 5 (1964), pp. 299–304.

Quine, Willard Van Orman
[1] *Notes on existence and necessity*. **Journal of philosophy**, vol. 40 (1943), pp. 113–127.
[2] *The problem of interpreting modal logic*. **The journal of symbolic logic**, vol. 12 (1947), pp. 43–48.
[3] ***World and object***. Cambridge, Mass.: The Technology Press of the M.I.T., and New York and London: John Wiley & Sons, 1960.

Ross, W. D. (Sir David)
[1] *Aristotle's Prior and Posterior Analytics*. Revised text with introduction and commentary by W. D. Ross. Oxford: At the Clarendon Press, 1949.

Smiley, Timothy
[1] Relative necessity. *The journal of symbolic logic*, vol. 28 (1963), pp. 113–134.

Sobociński, Boleslaw
[1] Note on a modal system of Feys–von Wright. *The journal of computing systems*, vol. 1 (1953), pp. 171–178.
[2] Remarks about axiomatizations of certain modal systems. *Notre Dame journal of formal logic*, vol. 5 (1964), pp. 71–80.

Sugihara, Takeo
[1] The number of modalities in T supplemented by the axiom CL^2pL^3p. *The journal of symbolic logic*, vol. 27 (1962), pp. 407–408.

Thomas, Ivo
[1] Ten modal models. *The journal of symbolic logic*, vol. 29 (1964), pp. 125–128.

von Wright, Georg Henrik
[1] Deontic logic. *Mind*, n.s. vol. 60 (1951), pp. 1–15.
[2] *An essay in modal logic*. Amsterdam: North-Holland Publishing Co., 1951.